Floating PV Plants

Floating PV Plants

Edited by

MARCO ROSA-CLOT
Professor of Physics
Scientific Director
Upsolar Floating Srl
Italy

GIUSEPPE MARCO TINA
Professor of Power Systems
Electrical, Electronic and Computer Engineering
University of Catania
Italy

ACADEMIC PRESS
An imprint of Elsevier

Academic Press is an imprint of Elsevier
125 London Wall, London EC2Y 5AS, United Kingdom
525 B Street, Suite 1650, San Diego, CA 92101, United States
50 Hampshire Street, 5th Floor, Cambridge, MA 02139, United States
The Boulevard, Langford Lane, Kidlington, Oxford OX5 1GB, United Kingdom

Notices
Knowledge and best practice in this field are constantly changing. As new research and experience broaden
our understanding, changes in research methods, professional practices, or medical treatment may become
necessary.

Practitioners and researchers must always rely on their own experience and knowledge in evaluating and using
any information, methods, compounds, or experiments described herein. In using such information or methods
they should be mindful of their own safety and the safety of others, including parties for whom they have a
professional responsibility.

To the fullest extent of the law, neither the Publisher nor the authors, contributors, or editors, assume any
liability for any injury and/or damage to persons or property as a matter of products liability, negligence or
otherwise, or from any use or operation of any methods, products, instructions, or ideas contained in the
material herein.

Library of Congress Cataloging-in-Publication Data
A catalog record for this book is available from the Library of Congress

British Library Cataloguing-in-Publication Data
A catalogue record for this book is available from the British Library

ISBN: 978-0-12-817061-8

For information on all Academic Press publications visit our website at
https://www.elsevier.com/books-and-journals

Publisher: Brian Romer
Acquisitions Editor: Lisa Reading
Editorial Project Manager: Ali Afzal-Khan
Production Project Manager: Sreejith Viswanathan
Cover Designer: Alan Studholme

Typeset by TNQ Technologies

List of Contributors

Raniero Cazzaniga, CTO, R&D
Koiné Multimedia
Pisa, Italy

Matt Folley, PhD, BSc
Senior Research Fellow
School of Natural and Built Environment
Queen's University Belfast
Belfast, Northern Ireland
United Kingdom

Jonathan Hancock, MEng (Hons), PhD, MICE, CEng
Solar Marine Energy Ltd. & Independent Engineer/ Consultant
Sparti, Greece

Giuseppe Marco Tina
Professor of Electric Energy Systems
University of Catania (UdC)
Catania, Italy

Marco Rosa-Clot
Professor of Physics
Scientific Director
Upsolar Floating Srl
Italy

Paolo Rosa-Clot
Independent Researcher
Pisa, Italy

Trevor Whittaker, PhD, BSc, FREng., FRINA, FICE, CEng
Professor
Faculty of Engineering and Applied Science
Queen's University Belfast
Belfast, Northern Ireland
United Kingdom

Contents

CHAPTER 1

Introduction

MARCO ROSA-CLOT • GIUSEPPE MARCO TINA

1. RENEWABLE ENERGY SOURCES: WHY FLOATING PV PLANTS?

Renewable energy sources (RESs) have been strongly increasing in the last decade with an overwhelming importance in the electricity sector. The electric sector represented 43% of energy demands in 2017, and this percentage will rise to 47% in the next 20 years [1]. At the same time, global warming and climate changes are the new challenges for mankind, and this crisis is mainly due to the burning of fossil fuels.

Globally RESs have registered an 8% yearly increase in the installed power in the last 10 years [1]. This increase is being driven by the sensational development of the photovoltaic (PV) sector which has registered a rate of growth of 45% and also by the wind sector (19%) with a more rapid growth for the offshore plants (33%).

The exponential growth of PV sector is slowing down, and if we compare the last decade we get Table 1.1 [2], where the yearly increase for the different RESs are given for the last two periods of 5 years.

It is quite evident that the rush of the PV sector is continuing because PV plants are simple, cheap, and easily and quickly installed. This increase is shown in synthesis in Fig. 1.1, where the four main components (hydroelectric, wind, photovoltaic, and biomass) are given. See Ref. [3] for the data.

However, the hydroelectric sector is even more important than what appears from Fig. 1.1 if we consider the electric energy production. Its contribution is more than 4 million GWh in 2017, that is, 73% of the energy produced by RES compared with the installed power which is only 58%. This is due to the very high capacity factor of hydroelectric plants.

Currently, the MWh produced in a year for any MW installed (capacity factor: CF usually given in hours) is on average 3294 MWh/year for any hydroelectric MW installed, and this value should be compared with solar PV (1146 hours) and wind farm (2183 hours); the only more efficient technology is the production with biomass which reaches 4635 hours and geothermal with 6326 hours. See Fig. 1.2.

Therefore, if we look to the renewable energy sector, even if the solar PV is quickly increasing and in 2017 it covered 17.7% of the renewable energy power, its contribution to the energy production is only 5%. In contrast, bioenergy which covers only 5% of the installed electric power reached 8.6% of the energy production, thanks to the large CF value. A plot is given in Fig. 1.3, where the large hydroelectric production has been omitted (4185 TWh in 2017) in order to highlight the new emerging technologies.

Notwithstanding the limited CF factor, PV is still expanding for several reasons:
- Simplicity and reliability
- Scalability
- Low costs
- Availability worldwide even near human settlement with limited environmental impact

There are, however, two main limits in the use of PV power source: the land use and the lack of incentives:
- **Land use:** The requirement for a large surface of land due to low PV panel efficiency (typically around 14%), this implies that a 1 MWp power station requires at least 15,000 m^2 of land, and this has a large environmental impact since the land cannot then be used for other purposes (agriculture, pasture, etc.).
- **Incentives:** The photovoltaic market was doped by very high incentives values. These were necessary for the start-up of the PV sector, but made the customers to use large land areas, which could have been exploited for other economic purposes. Since 2011, the incentives began to disappear at global level and, as a consequence, the PV market suffered a slowdown and the PV had to face the competition of other energy sources [1] (Fig. 1.4).

The effects of these two factors combined led to the contraction of the PV market in Europe and North America.

The lack of incentives has been partially overcome by the dumping of the PV modules, but the land disposal remains an important limit, especially in industrialized countries.

Floating PV Plants. https://doi.org/10.1016/B978-0-12-817061-8.00001-4
Copyright © 2020 Elsevier Inc. All rights reserved.

TABLE 1.1
Yearly Growth Rate for the Period 2007–12 and 2012–17.

	Rate of Growth	Rate of Growth
	5 years 2007–12	5 years 2012–17
TECHNOLOGY		
Total renewable energy	7.7%	8.6%
Renewable hydropower	3.4%	3.2%
Wind	23.8%	14.0%
Solar photovoltaic	61.9%	31.7%
Bioenergy	9.1%	6.9%

The wind energy sector has partially solved the problem of land occupancy. The production of huge wind turbines triggered a great expansion in this sector, which reached 5% of the total worldwide production in 2017. The availability of offshore technology contributed to this trend, and 24% of wind power installed in Europe in 2015 was constituted by offshore wind turbines.

Actually if we wish to tilt the balance in favor of solar energy production and fully develop solar potentialities, the availability of large surfaces not far away from urban settlements should be granted.

A technology which can avoid land use is the floating PV (FPV), which has had a strong increase in the last few years with the installation of PV plants on free water surfaces, the exploitation of existing basins, and in some cases it has been coupled to hydroelectric, thanks to the easy integration of the two technologies. See Ref. [4].

Up to now the limit of this technology was the high prices with respect to land-based PV. A technology which can lower the price of floating PV is a market breakthrough highly innovative solution that goes beyond the state of the art of the existing solutions.

2. THE CURRENT SITUATION OF FPV

It is impossible to give a detailed analysis of the many small (less than 1 MWp) PV floating plants built in the last 10 years. The plot here below is based on data taken from the World Bank Group [5]. About 585 MWp is the cumulative value of FPV in 2017, whereas 1100 MWp is a preliminary value for the 2018. The yellow bar for 2019 is an extrapolation assuming the trend of an exponential increase with an annual growth rate of 140%. See Fig. 1.5, where a logarithmic plot is given together with an interpolating exponential fit. Other authors confirm the quick increase of the sector [6].

This trend has been outlined in a recent paper on PV magazine [7] where the authors speaks of an "unstoppable tide" and forecast 13 GW of new floating PV additions in the world in the next 5 years (Fig. 1.6).

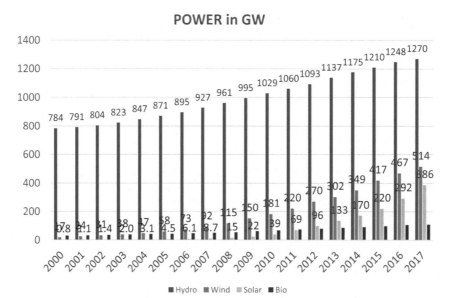

FIG. 1.1 Worldwide power in GW for the main renewable energy sources.

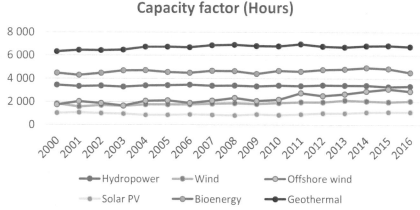

FIG. 1.2 Capacity factor of the different renewable energy sources.

3. FLOATING PV PLANTS: WHERE?

As mentioned above, the large-scale deployment of PV energy entails the use of a significant amount of land. In the United States, the capacity-weighted average land use for large PV plants ranges from 0.5 to 0.7 MWp/ha and the land availability is related to the concept of geographic potential. This concept can be extended to water surfaces. In this case, technical problems are very different from land-based PV plants, and it is possible to arrange the modules more compactly increasing the previous values to more than 1–1.5 MWp/ha.

Furthermore, it must be observed that wherever human settlements are built, water is also present. This can be found in a variety of forms such as lakes, seas, large artificial basins built for various purposes (water storage, irrigation, or civil use), wastewater treatment, hydroelectric basins, abandoned mines, etc. These very large existing surfaces suggest a very simple solution to the problem of power/surface limitations: they could be used to install FPV plants.

How large are these surfaces? Can they account for a substantial expansion of the PV sector and increase its contribution to RES? A simple analysis of the available water surfaces shows that very large freshwater basins are available everywhere.

For example, in Sicily, one of the driest regions in Italy, there are over 75 km^2 of large freshwater basins. Even more can be found by considering small irrigation basins and water reservoirs suitable for FPV plants because PV installations favor water saving and water quality control.

Table 1.2 shows the values of freshwater surfaces, the installable PV power (the so-called technical power potential) if only 1% of these surfaces is used, and the corresponding potential energy production (technical energy potential) for extended regions worldwide: tropic, temperate, and cold zones [8] (see Chapter 3).

It should be noted that, notwithstanding only 1% of the surfaces is taken into account, the potential energy which could be produced on freshwater basins is 5988 TWh and which would cover about the 25% of the entire world production of electric energy, which in 2014 was 23,816 TWh [9,10].

This potential is enormous: even considering that many large basins are not easily or immediately exploitable; the numbers are impressive and clearly indicate the advantages that would accrue from the exploitation of these untapped resources.

The extension of the PV floating solution to the sea (to near-shore and offshore plants) multiplies the potential of water surfaces. Obviously, the simplest solutions cannot be found in the open ocean where very large waves have a destructive impact and where the distances between the floating plant, the end users, and the interconnected electricity grid heavily increase the costs and the technical challenges. Rather, the idea is to build floating structures not too distant from the coastline and to choose locations with a natural (or artificial) limit to wave strength.

4. ADVANTAGES OF FPV

FPV plants open up new opportunities that have not been fully explored. The main advantages can be summarized in the items below:

1. **Strong reduction of land occupancy.** The main advantage of floating or submerged PV plants is that they do not take up any land, except the limited surfaces necessary for electric cabinet. FPV plants are not merely more economical than land-based

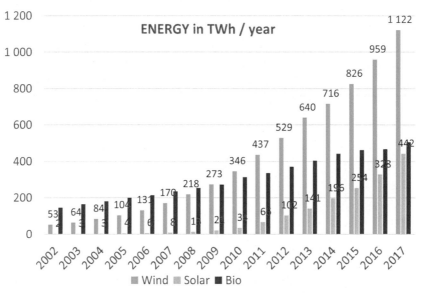

FIG. 1.3 Worldwide energy production in GWh for the main renewable energy sources.

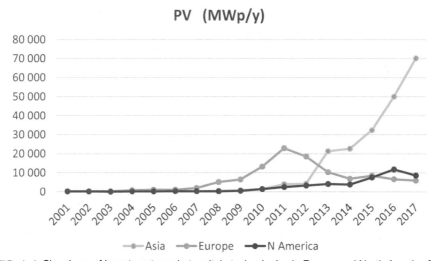

FIG. 1.4 Slowdown of investment on photovoltaic technologies in Europe and North America [1].

plants, but they provide mainly, and above all, a way to avoid competing with agricultural or green zones. Also, unlike land-based PV plants, floating or submerged plants have a more limited impact on the landscape.

2. **Installation and decommissioning.** FPV plants are more compact than land-based plants, their management is simpler and their construction and decommissioning straightforward. The main point is that no fixed structures exist, and the mooring of

floating systems can be carried out in a totally reversible way, unlike the foundations used for a land-based plant.

3. **Water saving and water quality.** The partial coverage of basins has additional benefits such as the reduction of water evaporation. This result depends on climate conditions and on the percentage of the covered surface. In arid climates (such as Central Australia or Sael region), this is an important advantage since more than 80% of the evaporation

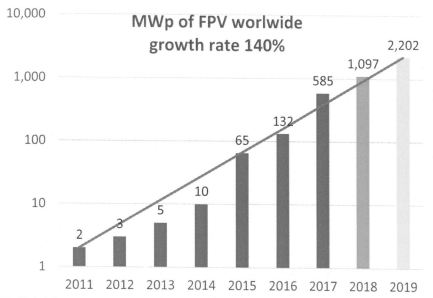

FIG. 1.5 Global floating photovoltaic installation up to 2017 (preliminary value for 2018 in blue and 2019 forecast in yellow).

of the covered surface is saved and this means more than 20,000 m^3/year/ha (a very useful feature, especially if the basin is used for irrigation purposes).

4. **Cooling and tracking.** The floating structure allows the implementation of a simple and cheap cooling and tracking mechanism. A large floating platform can easily rotate and can perform a vertical axis tracking: this can be done without wasting energy and without the need for any complex mechanical apparatus, which is needed in land-based PV plants. Moreover, a floating PV plant equipped with a tracking system has a limited additional cost, while the gain in energy can range from 15% to 25%. In several cases, the technical effort necessary to implement this solution gives a sizable reduction of the final kWh cost.

5. **Hybrid system and in particular coupling to the hydroelectric power plants (HPP).** FPV plants can advantageously integrate other RES technology. Solar energy is partially anticorrelated to the wind energy which, especially in North area is related to bad weather condition. So the same grid can be used to support large offshore wind farm and FPV plants. However, the main advantages come from the coupling with HPP where the natural storage

possibilities allow to increase the capacity factor of the coupled systems.

6. **Environment control.** A parallel advantage is the containment of the problem of algae bloom, which is especially serious in industrialized countries. The partial coverage of the basins and the reduction of light on biological fouling just below the surface, together with active systems, can solve this problem.

7. **Synergy with fishing.** Several projects have been presented which couple the FPV to activities related to fish or shrimp farms, mainly in China and in the Asia South East [11].

8. **Limiting greenhouse effect.** Due to the strong reduction of albedo effects on PV plants on land (or on roofs), the radiation energy balance is negative and contributes to the global warming. Conversely floating plants leave the radiation balance unaltered, and this should be considered an important advantage especially if the installed power is destined to increase considerably.

9. **Reduction of specific energy cost.** This is a very important item, perhaps the crucial one. The evolution of the FPV technology brought the costs of the FPV plants below that of standard PV especially in the tropical regions where the land management and the maintenance of land-based plants are very

FIG. 1.6 40-MW China plant (Sungrow Power).

TABLE 1.2
Technical Photovoltaic Potential for Climate Zones.

	Surfaces, km²	Technical Power Potential, GWp	Technical Energy Potential, TWh/year
Tropical zone	1,448,031	1875	2352
Temperate zone	1,386,202	1677	1922
Cold zone	1,611,663	1715	1714
	4,445,896	**5267**	**5988**

expensive, and currently the kWh cost ranges between 3 and 6 cents of $ depending on the local radiation [10–12].

5. BOOK PLAN

The FPV has been called the third pillar of photovoltaic sector after the land and the roof plants, but the evolution is only at the very beginning and many new aspects will quickly expand in the next few years. In this book, we analyze the aspects which are more important for practical purposes.

- The present status on the future trends of the FPV plants are analyzed in Chapter 2
- Chapter 3 is dedicated to the location and geographic potential evaluation.
- Structural elements of the floating platforms are analyzed in Chapter 4 where some innovative solutions are also discussed and is integrated by

Chapter 5 where wind load and waves impact are studied and discussed with the support of numerical simulations.

- Chapters 6 and 7 are devoted to two facilities which can be easily and cheaply implemented on FPV plants: cooling and tracking systems.
- Chapter 8 is dedicated to the experimental tests. These aspects are at the very beginning and there are few data published, but some systematic tests have been performed and will be quoted and analyzed.
- Chapter 9 studies the hydroelectric integration in detail. As the hydroelectric energy is the more important contribution in the RES, it is crucial to discuss the natural coupling of the FPV technology with the hydroelectric sector.
- Offshore FPV systems have a limited application, but due to the very large seawater surfaces available, studying them can be crucial, and they will be further analyzed in Chapter 10.
- Environmental problems are discussed in Chapter 11 where impact on the landscape and on the water quality is discussed. Particular emphasis is given to the lifetime and recycling of the plant as well as to the global warming problem.
- Lastly, Chapter 12 deals with economic problems and discusses the possible business plans for the future FPV, as well as the trend of the sector.

REFERENCES

[1] BP Statistical Review of World Energy, BP Energy Outlook, London, 2017.
[2] IRENA, Renewable Capacity Statistics 2018, International Renewable Energy Agency, Abu Dhabi, 2018.

[3] Renewables 2017 Global Status Report, REN21 Steering Committee, 2018.

[4] M. Rosa-Clot, M. Tina, Submerged and Floating Photovoltaic Systems, Academic Press, 2017.

[5] Where Sun Meets Water: Floating Solar Market Report, World Bank Group and SERIS, Singapore, 2018.

[6] Dang Anh Thi Nguyen, The Evolution of Floating Solar Photovoltaics, July 16, 2017.

[7] B. Beetz, "PV MAgazine: 14 PV trends for 2019," [Online]. Available: https://www.pv-magazine.com/2018/12/31/14-pv-trends-for-2019/. [Accessed 2 January 2019].

[8] C.I.A. US, The World Facebook, 2017 [Online]. Available, https://www.cia.gov/library/publications/the-world-factbook/.

[9] I.E. International Energy Agency, Key World Energy Statistics, IEA, 2016.

[10] M. Rosa-Clot, G.M. Tina, Submerged and Floating Photovoltaic Systems, Modelling, Design, Case Studies., Elsevier, Academic Press, London, 2017.

[11] L. Wasthage, Optimization of Floating PV Systems; Case Study for a Shrimp Farm in Thailand, Mälardalen University, 2017.

[12] Lazard, Lazard's Levelized Costs of Energy Analysis — Version 10.0, December 2016.

CHAPTER 2

Current Status of FPV and Trends

MARCO ROSA-CLOT • GIUSEPPE MARCO TINA

1. INTRODUCTION

Many research centers and economic operators claim that a deep and huge change is underway in the world of energy production/consumption.

Among the many analyses of the phenomenon (see, for example, [1−6]), we quote the synthetic pictures published recently by DNV which summarize the situation [7]; from now to 2050 several deep changes will happen as shown in Fig. 2.1.

The plot highlights the strong dumping of fossil fuels, the slow decline of nuclear power, the emergence of wind and solar photovoltaics (PV), and the slight increase of the hydroelectric. In 2050, all the renewable energy sources (RES) together will account for almost 80% of the electric energy production with a dominance of more than one-third due to the solar PV. See Fig. 2.2.

All these analyses are perhaps too optimistic in some cases and related to specific development models (it is not sure if by 2050 more than 1 billion cars will be fully electrically driven [2]), but they converge toward few concepts:

1. Electric production will rise to more than 50% of the final energy consumption
2. Electric production by fossil fuel and nuclear will drop to 20% (less than half the current value)
3. The RES will be more than 80% of the full electric energy production with a dominant contribution from solar and from wind sources which will compete for first place in electric energy production.

These analyses, however, do not take into account the emergence of the new floating PV (FPV) technology which covers currently less than 0.1% of the electric energy production.

Actually FPV is emerging as a new cheap and robust technology, but the most important aspect is that it can be integrated to many existing structures without wasting land and without relevant preliminary works [8].

There is a difference between FPV technology and other innovative energy sectors which experience a strong innovation revolution. FPV technology profits from the dramatic drop of the PV prices and offers a new simple solution. From the very beginning of the research in this sector (first work goes back to 2008 [9,10]), the surprising and exciting aspect was the multitude of approaches and of imitators. This means that the idea was mature, simple, and easily accessible and that many research centers, universities, and investors began to develop it in parallel.

The FPV in fact overcomes the main limit of the PV technology which lies in the extensive surfaces required and in the limited energy yield per m². More than 1−1.5 km² is necessary to get 100 MWp with land-based PV, and for this reason projects have been developed for desert areas. These, however, have to face the problem of a complex management and of the grid connection, and so huge projects like Desertec (a large-scale project supported by a foundation of the same name and the consortium Dii Desertec industrial initiative created in Germany) are focused to the use of large Sahara surfaces generate a lot of complex problems.

The FPV on the contrary is available everywhere where human activities and installments are present. In general, in these locations there are also water surfaces (either for wastewater treatment or for irrigation purpose or just only natural), and this means that large floating PV plants can be installed without the need for new structures, but simply using existing ones at a very low price with cost per kWp which are below $800.

This is not the case for other technologies which have to fight with intrinsic difficulties as, for example, the wave energy sector where hundredth of projects face the big problem of efficiency/cost rate, or with technical complex problems like in the storage energy sector and in the electric driven cars.

For these reasons—simplicity and low threshold entrance—FPV is quickly increasing at an impressive rate, 133% per year (data of the last 10 years), which promises an explosion of this technology around the world but mainly in the equatorial area where solar radiation is more intense. See Chapter 1.

Floating PV Plants. https://doi.org/10.1016/B978-0-12-817061-8.00002-6

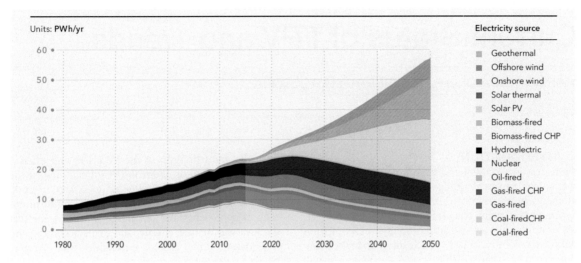

FIG. 2.1 World electricity generation by source DNV. Scale in PWh/yr (1000 TWh/yr).

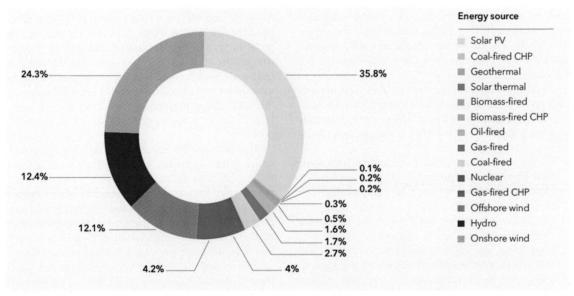

FIG. 2.2 Global electric energy production in 2050 per energy sources [2].

Many plants have been installed since 2017, and many projects have been launched by private companies and also by governmental organizations around the world. The worldwide situation is in very rapid evolution, and for this reason it is neither possible to give a precise update of the existing floating nor a complete analysis of the perspective of the sector, as they are both in quick evolution.

It is hard to overestimate this potential (see Chapter 3 for an analysis of the geographic potential), but the availability of already existing basins is very large. Furthermore, coupling to hydroelectric basins (see Chapter 7) with a hybrid production, part by hydroelectric turbine and part by floating PV, is a further argument which justifies the forecast of a strong increase of this sector in the next few years.

A recent wide report of the World Bank with its Energy Sector Management Assistance Program (ESMAP) in collaboration with SERIS [11] put together the main aspects of the sector and confirmed the perspective of a strong increase of floating PV energy production [4]. We will refer mainly to this work with the consciousness that the market moves too quickly to be fixed in some 2018−19 data table and that the FPV deployment appears likely to accelerate as the technologies mature, opening up a new frontier in the global expansion of renewable energy and bringing opportunities to a wide range of countries and markets.

2. EVOLUTION OF THE MARKET UP TO 2018 AND COMPANIES ACTIVE IN THE SECTOR

The global installed FPV capacity exceeded 1.3 GWp as of December 2018 and has been growing exponentially since 2010.

As shown in Fig. 1.5 of Chapter 1, there has been an exponential increase in the last 10 years with the total number of FPV plants more than doubling each year.

So it is rather useless to give a list of the main plants in the world: every month there is an announcement about the largest floating PV and longer lists are published which are always incomplete, thus confirming the popularity of this technology. The more recent record seems to be the 150-MWp FPV plant in the coal mining subsidence area, Huainan City (Panji—China

Three Gorges New Energy) managed by a consortium of China industries (Beijing NorthMan, Zhongya, Hefei Jintech New Energy Co. Ltd., Anhui ZNZC New Energy Co. Ltd., CJ Institute China). However, this record will soon be overtaken by other investors.

The main question is if this trend will continue in the next few years with the forecast shown in Fig. 2.3, which would imply that the market of the coming years will be conditioned by the FPV phenomenon, or if this is a burst which is just a short-time anomaly in the quick increase of the renewable energy sector.

Currently, the share of FPV plants around the world between different areas is shown in Fig. 2.4.

China dominates the market which huge investment, but Japan and East Asia (mainly South Korea)

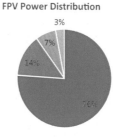

FIG. 2.4 Share of floating photovoltaic (FPV) plants between different world areas [5].

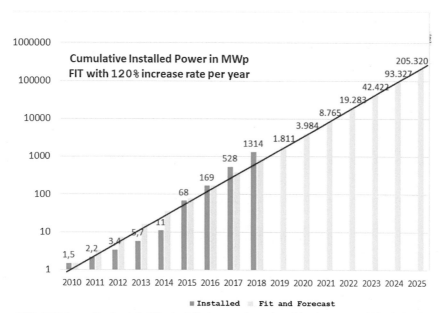

FIG. 2.3 Logarithmic plot of the installed power (green) and forecast up to 2025 (yellow).

also have an important role. The absence of India is due to the fact that this plot goes back to 2017, but now India and many other countries in the equatorial region are doing big investments in this sector.

In Fig. 2.5, an analysis of the installed PV has been done with regard to the dimensions of the plants. It is quite evident that the distribution peaks toward the large size plants, and this is actually the main characteristic of this technology which is scalable but favors large plants in already existing basins.

These analyses suggest that the development of the FPV sector will be dominated by large plants and by large governments programs. Following ESMAP analysis and integrating it with direct information, 40 governments have programs of investment, feed-in tariffs, research centers, or companies active in the FPV sector. The list below is simply an index of these countries with government programs which globally amount to more than 500 GWp for 2030, or with companies investing in the FPV sector. This list, however, is in quick evolution as well as government programs of investment which will not always be realized (recently India announced a program of 100 GWp for the 2022 with a suggestion of a 1000 GW potential for FPV systems [12]) (Table 2.1).

This list is, of course, not exhaustive, and furthermore, beyond the government programs, there are many research centers which are active in the FPV sector such as Mirarco, SERIS, Fraunhofer, and many others with a relevant number of universities which is so large that it is impossible to quote any of them.

TABLE 2.1
Nonexhaustive List of Countries With Government Programs or Private Companies in the Floating Photovoltaic Sector.

Albania	Denmark	Laos Dem. Republic	Sri Lanka
Australia	Dubai	Malaysia	Taiwan, China
Azerbaijan	France	Norway	Thailand
Bangladesh	Ghana	Pakistan	The Netherlands
Belgium	India	Panama	Tunisia
Brazil	Indonesia	Portugal	UAE
Cambodia	Iran	Republic of Korea	Ukraine
Canada	Italy	Republic of Maldives	United Kingdom
China	Jamaica	Seychelles	United States
Colombia	Japan	Singapore	Vietnam

In the same way the number of companies working on the FPV sector is exploding, and here we give a provisional list of 45 companies which are installing floating PV with different technologies, but all based on the concept of floating rafts with superposed PV modules (Table 2.2).

This explosion of interest is motivated by several reasons but mainly by the following factors:

- FPV plants are cheap and can lower the kwh price below that of land-based PV: we are speaking of cost ranging from 2 to 5 cent of $ per kWh depending on the local radiation yield.
- FPV plants are easily installed, with very limited environmental impact, and are easily integrated to existing structures. So the cost threshold is very low, and it is not necessary to invest a lot of money and time to get onto the market.
- Decommissioning is simple, and the area interested by the plant is returned to initial conditions without any specific interventions.

But what is the real potential and trend of the sector?

3. FUTURE TRENDS AND PERSPECTIVES

The contribution to the total worldwide electric energy production is given in TWh per year in Fig. 2.1 with a projection to 2050.

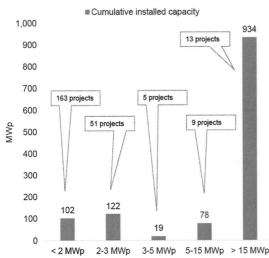

FIG. 2.5 Floating photovoltaic plants distribution by power [5].

TABLE 2.2
Nonexhaustive List of 45 Companies Supplying Floating Photovoltaic System (August 2019).

Company Name	Country	Website
1. 4C Solar	USA	https://www.4csolar.com/
2. Adtech	India	https://adtechindia.com/solar-energy/
3. Ciel and Terre International	France	https://www.ciel-et-terre.net/
4. ENEL GP	Italy	https://www.enelgreenpower.com/country-italy
5. ENI	Italy	www.eni.com
6. Floating Solar	The Netherlands	https://floatingsolar.nl/en
7. HDB	Singapore	https://www.hdb.gov.sg/cs/infoweb/homepage
8. Isifloating	Spain	https://www.isifloating.com/
9. ISIGENERE	Spain	https://isifloatingcom.wordpress.com/
10. Jintech New Energy	China	http://www.jnnewenergy.com/
11. Kyocera	Japan	https://global.kyocera.com/
12. Koiné	Italy	http://www.koinemultimedia.eu/wp/
13. Kyoraku Co.	East Asia	http://www.krk.co.jp/
14. LG CNS	Korea, Rep.	http://lgcns.co.kr/
15. LS Industrial Systems	Korea, Rep.	http://www.lsis.com/ko/
16. NorthMan Energy Tech	China	https://netsolar.solarbe.com/
17. NRG Energia Italy	Worldwide	http://www.nrg-energia.it/index-en.html
18. Ocean Sun	Norway	http://oceansun.no/
19. Oceans of Energy	The Netherlands	https://oceansofenergy.blue/
20. Phoenix Solar	Germany	http://www.phoenixsolar-group.com/
21. ProFloating	The Netherlands	https://profloating.eu/en/
22. REC	Germany	https://www.recgroup.com/en
23. Scatec Solar	Sweden	https://scatecsolar.com/
24. SCG Chemicals	Thailand	https://www.scgchemicals.com/en
25. Scotra Co.	Korea	http://www.scotra.co.kr/en/
26. Sharp Solar	Japan	http://www.sharp-world.com/solar/en/
27. Solar Energy Corporation	India	http://seci.co.in/
28. Solar Marine Energy	Ireland	http://solarmarinenergy.com/
29. Solargy	USA	https://solergy.co/
30. Solaris Synergy	Israel	http://www.solaris-synergy.com/
31. SolarisFloat	Portugal	https://www.solarisfloat.com/
32. Statkraft	Norway	https://www.statkraft.com/
33. Sumitomo Mitsui Co.	Japan	https://pv-float.com/english/
34. Sunhome	China	https://www.sunhomepv.com/
35. Sun Rise E&T Corporation	Taiwan	http://www.srise.com.tw/v2/
36. Sunengy	Australia	http://sunengy.com/
37. Sunfloat	Netherlands	http://www.sunfloat.com/

Continued

TABLE 2.2
Nonexhaustive List of 45 Companies Supplying Floating Photovoltaic System (August 2019).—cont'd

Company Name	Country	Website
38. Sungrow	China	https://en.sungrowpower.com/
39. SunSeap	Singapore	https://www.sunseap.com/SG/
40. Swimsol	Austria	https://swimsol.com/
41. Takiron Engineering	Japan	https://www.takiron.co.jp/english/
42. Upsolar floating	Italy	www.floatingupsolar.com
43. Vikram Solar	India	https://www.vikramsolar.com/
44. Xiamen Mibet New Energy Co.	China	https://www.mbt-energy.com/
45. Yellow Tropus	India	http://www.yellow.org.in/

TABLE 2.3
Value of Photovoltaics Installed (in MWp) for the Main Countries and Forecast to 2023.

	2018 MWp	2023 MW	Growth %		2018 MWp	2023 MW	Growth %
China	175.131	448.131	21%	France	8.920	22.259	20%
India	27.347	116.106	34%	Saudi Arabia	19	11.412	260%
United States	62.127	132.426	16%	Brazil	2.346	12.505	40%
Australia	12.560	45.236	29%	Italy	19.877	29.498	8%
Germany	45.920	72.611	10%	Taiwan	2.739	12.074	35%
Japan	55.851	82.351	8%	Pakistan	1.720	8.381	37%
Spain	5.915	23.367	32%	Ukraine	2.004	7.963	32%
South Korea	7.742	24.768	26%	Turkey	5.062	10.562	16%
The Netherlands	4.181	20.059	37%	UAE	720	6.132	53%
Mexico	3.580	19.010	40%	Egypt	661	5.023	50%
				Total	**444.422**	**1.109.874**	**20%**

This behavior is generally accepted by other research centers and can be summarized, as far as RES is concerned, in Table 2.3 for the sectors growth rate [12]. These values have been obtained by extrapolation of the exponential growth of the five sectors in the last 5 years, and we can assume that this growth will last in the next future for 5—10 years.

This of course cannot be true for sectors with a very high yearly increase such as floating PV sector. Actually applying the 133% growth rate we should produce in 2030 all the electric energy by FPV which is a nonsense.

So some of the values quoted in Fig. 2.6 change from one research center to another, but a comparison with other forecasts shows changes only in the second digit of the values quoted. See, for example, Ref. [13].

Particular care is required for the PV sector where the very strong growth rate is questionable. In particular,

Solar Power Europe [14] gives a detailed, but not complete, analysis as that shown in Table 2.3 for the solar PV sector which suggests a more limited growth for the PV sector.

In order to develop the analysis, we will assume the growth rate shown in Fig. 2.6 with a correction for the PV growth which will be assumed to be 25% per year a value just in between the trend of last 5 years and the forecast suggested in Table 2.3.

Collecting these results and historical series up to 2018, we plot the behavior of the electric energy production for sector in Fig. 2.7; results are given in TWh and the total electric produced energy is given by the black line.

- Due to the generalized exponential increase of all the quantities under analysis the logarithmic plot is compulsory. Furthermore, the analysis has been stopped to 2030 since longer-term forecasts are

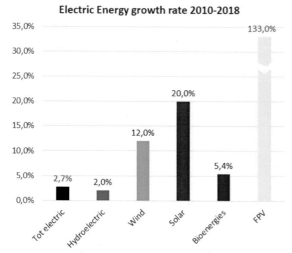

Electric Energy growth rate 2010-2018

FIG. 2.6 Growth rate of renewable energy source energy sector up to 2030. The value for floating photovoltaic (FPV) has been calculated using the last 10 years.

affected by too strong exogenous factors which could not be beyond control.

- Stars represent the measured value and are given up to 2018.
- The dash-dotted lines represent the possible behavior up to 2025 with growth rates which have been discussed above.

- The value for FPV has been calculated using the values for the last 10 years and extrapolating with the same growth rate. In order to compare the energy production an average value of 1200 kWh/y/kWp has been assumed.
- The FPV sector (cyan line) is critical. A simple extrapolation of the last 5 years trend would bring to clearly untenable forecast results, and for this reason a second dash-dotted line with a lower growth rate of 120% has been assumed.

The plot puts in evidence the very quick increase of the PV sector which will approach the energy production by the wind sector in the 2025 and will overcome it in the 2028.

The reasons for this overtaking are simple: solar energy, thanks to the dramatic cost drop of PV modules and inverters, is very cheap and very easy to install at variance with hydroelectric which requires big structures and has a very strong impact on the environment.

The two "light" competing technologies, wind and biofuel, have a more limited growth rate for two reasons: wind requires a quite complex mechanical maintenance, whereas the very important sector of bioenergy is still looking for the optimal solution and is intrinsically limited by the waste disposal of burnable organic material.

The main limit of PV technologies is land availability, and this problem seems to be solved by the FPV technologies.

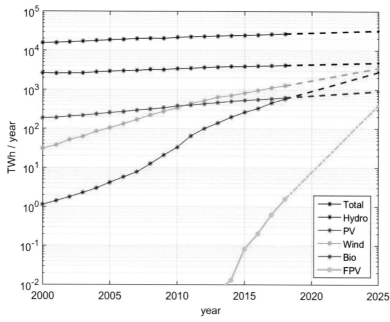

FIG. 2.7 Logarithmic plot of the main contributions to the electricity production. *FPV*, floating photovoltaic; *PV*, photovoltaics.

Some factors work in favor and others against this behavior.

In favor of this quick increase, there are a few elements:

1. The improvement of the technology, which is going to become more robust and will cost less. Actually, even if simple, and with a low entry threshold, the FPV technology is at its very beginning, and we think that important reductions of raft price and of logistic cost are in progress.
2. The understanding from energy operators and from the market that integration with hydroelectric power plants is very advantageous (and this could perhaps orientate in a different way investments in this sector, slightly reducing the growth rate of the hydroelectric sector).
3. The penetration of the technology in the equatorial area and the very low price of the kWh produced in these regions (below 3 cents of $) will help emerging countries in their development and will offer the opportunity of a control of natural resources with a minimal environmental impact.
4. The extension of the technology to near offshore and to salty shallow water will further multiply the possibilities of installation of these energy systems.

These elements which favor a strong increase of the FPV sector are contrasted by two boundary conditions:

1. The limiting resources and the limited PV global production. An expansion of the FPV sector will strongly interfere with the development of the PV sector itself increasing it in some way but with a constraint due to the full capacity of the market to supply the necessary technical and economical resources.
2. The necessity of cheap and efficient storage systems, which are lacking up to now. Actually we must not forget that solar energy is intermittent and that its availability seldom exceeds 1500 hours per year. This is a very weak point which can be only compensated by a very low cost of the energy stored, below 100 $/kWh.

Given all these elements, we can modify our forecast and rather than extrapolate a very strong increase of the FPV sector, we introduce corrections to this trend.

First of all, we will assume that the penetration of FPV is biased by a general constraint: the full PV market can be influenced by the contribution of FPV but only to a certain extent which limits the strong growth rate at a two-digit value. At the same time, the strong expansion of the FPV will partially reduce the growth rate of the standard PV sector.

Second, we suggest that the hydroelectric sector will divert part of the investment (10%) in this sector toward the FPV coupled to the hydroelectric sector.

Other minor corrections can be viewed, but these are beyond the purpose of this first global analysis.

These corrections limit to 227 TWh the FPV energy produced in the 2025 (189 GWp of installed power to be compared with the forecast of 205 given in Fig. 2.3). These values have to be compared with 2788 TWh forecast for the standard PV market and suggest that in 2025 the FPV market will be about 7% of the full PV market.

This behavior gives a global idea of the market evolution but deserves further deeper analysis.

In Fig. 2.8, the general trend is given in a logarithmic plot. The dotted line represents the behavior of the RES components taking into account the important development of the FPV sector.

Three phenomena are important:

- Energy produced by wind and by PV power plant will be more than the hydroelectric energy production by the 2028.
- In the same period, the PV energy production will reach the same as the one with wind.
- The FPV TWh will be in the 2030, a 17% of the full PV energy production

Table 2.4 collects in an explicit form the information of the curves in Fig. 2.8.

The suggested model shows an increase of the renewable energy rate with respect to the total energy production with a rate which rises to 61% in the 2030.

The dynamic of hydroelectric sector is slightly reduced due to the diversion of capital investment in the hybrid FPV-HPP plants, and the standard traditional PV sector (mainly ground-based plant and roof plants) is diminished due to the expansion of the FPV sector. Together the full PV (standard PV + FPV) increases its expansion and overtake the wind sector energy production in the 2028.

The FPV sector records a reduction of the growth rate to a two-digit value (46%) but remains the more performant component of the renewable energy sources.

Finally, the plot in Fig. 2.9 gives a plot in linear scale of the main components of RES and highlights the importance of the emerging FPV technology.

One point, which has been neglected in this analysis, is the limited capacity factor of solar PV which seldom exceeds 1500 hours and which is lower than wind and hydro capacity factor. This is a real limit to the PV and FPV sector development and could be overcome only by a very efficient storage system with a cost below 100 $ per kWh stored.

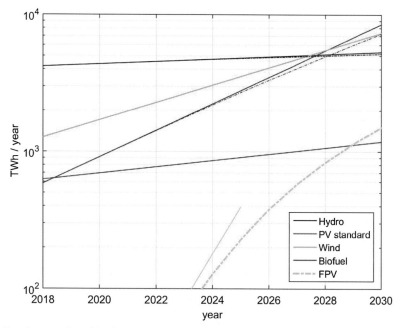

FIG. 2.8 Trends correction of the forecast up to 2030. Dash-dotted lines are the corrections to the naïve trend extrapolation of Fig. 2.7. *FPV*, floating photovoltaic; *PV*, photovoltaics.

TABLE 2.4
Growth of Energy Production 2018–30.

Electric Energy Production	2018 TWh	2025 TWh	Growth Rate	2030 TWh	Growth Rate
Total electric energy	26615	32143	2.7%	36681	2.7%
hydro	4193	4820	2.0%	5166	1.4%
Wind	1270	3536	15.8%	7349	15.8%
Bio	626	907	5.4%	1181	5.4%
PV standard	585	2788	25.0%	7267	21.1%
FPV	1.6	227	103.4%	1499	45.9%
Total PV	5867	3015	26.3%	8766	23.8%
Total RES	6676	12278	9.1%	22462	12.8%
RES/Tot. Electric Energy	**25%**	**38%**		**61%**	

FPV, floating photovoltaic; *PV*, photovoltaics; *RES*, renewable energy source.

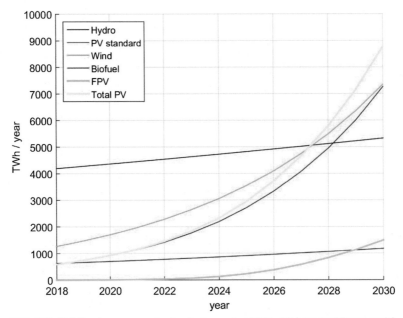

FIG. 2.9 Details of energy production in the period 2018—30 (standard linear scale).

A lot of investments are underway, and this is a really promising sector which could further speed up the transition to a fully green energy production.

REFERENCES

[1] Bloomberg [Online]. Available: https://about.bnef.com/new-energy-outlook/.

[2] BP Statistical Review of World Energy, BP Energy Outlook, London, 2017.

[3] IEA, Key World Energy Statistics, IEA, 2016.

[4] IRENA, Renewable Capacity Statistics 2018, International Renewable Energy Agency, Abu Dhabi, 2018.

[5] O. Knight, Where Sun Meet Water, in Norwep International Solar Day, Oslo, Norway, Oslo, 2019.

[6] A. Jager-Waldau, PV Status Report 2018, 2018. JRC Science for Policy Report, Luxemburg.

[7] Energy Transition Outlook : Renewables, Power and Energy Use Forecast to 2050, DNVGL Energy Headquarters, Utrecht.

[8] N.D.A. Thi, The Evolution of Floating Solar Photovoltaics, July 16, 2017.

[9] M. Rosa-Clot, P. Rosa-Clot, Support and method for increasing the efficiency of solar cells by immersion, Italy Patent PI2008A000088, 8 September 2008.

[10] Y. Ueda, T. Sakurai, S. Tatebe, A. Itoh, K. Kurokawa, Performance analysis of PV systems on the water, in: 23rd European Photovoltai Solar Energy Conference, Valentia, 2008.

[11] Where Sun Meets Water: Floating Solar Market Report, World Bank Group and SERIS, Singapore, 2019.

[12] A. Sharma, Floating Solar PV potential in large reservoirs in India, International Journal for Innovative Research 2 (11) (2016) 97—101.

[13] OECD-IEA and IRENA, Perspective for the Energy Transition, IEA Publication, IRENA Publications, German, 2017.

[14] W. Hemetsberger, M. Schmela, Global Market Outlook for Solar Power 2018-2023, 2019. Brussels.

CHAPTER 3

Geographic Potential

MARCO ROSA-CLOT • GIUSEPPE MARCO TINA

1. SOLAR PV: WHERE?

Several elements contribute to determine the optimal location of a solar plant. The most important are the solar radiation intensity and the wide surfaces availability.

In this chapter, we will analyze mainly these two aspects with specific attention to the water surface where floating photovoltaic (FPV) plants can be installed.

Solar radiation characteristics are widely analyzed in many papers and books [1,2], so we do not enter into the general aspects of this specific field, but we focus the attention on what is useful for calculating the energy yield of a FPV plant.

The main points of interest are the solar radiation components:
- direct (defined by the solar zenith angle, θ_z, and the solar azimuth angle γ_s),
- diffuse due mainly to scattering on the atmosphere and clouds,
- reflected by earth surface; this last one is called albedo.

In Fig. 3.1, a scheme of the components of solar radiation that reach the ground and a PV module are reported [3]. The global radiation that strikes the PV module, G_{PV}, is the sum of three components: $G_{PV,b}$ (beam radiation), $G_{PV,d}$ (diffuse radiation), and $G_{PV,r}$ (reflected radiation).

Diffuse radiation can be more than 70% of the solar radiation on cloudy days and is seldom below 30% even on clear sky days. This variability strongly affects the harvesting of solar radiation and restricts the possibility to use reflectors or concentration systems.

In Section 2, results for the energy harvesting will be analyzed for three possible FPV plants:
- Fixed floating systems
- Systems with horizontal axis tracking (HAT)
- Systems with vertical axis tracking (VAT)

In Section 3, albedo effect will be analyzed in detail with reference also to the interesting emerging technology of the bifacial PV modules.

The geographic potential will be studied in Section 4 and in Section 5 with particular attention to the new

possibilities suggested by the recent works about FPV systems: in particular fresh and salty water surfaces will be discussed in detail as well as wastewater basins. It is interesting to exploit the fact that one of the main advantages of FPV plants lies in the possibility to integrate them into already existing water basins built for civil, industrial, or agricultural use. A specific analysis of the hybrid coupling between hydroelectric power plant and FPV plants will be given in Chapter 8.

2. THE SOLAR RADIATION HARVESTING

The solar radiation depends mainly on the solar constant and on the latitude. Solar constant varies along the year (in winter the earth reaches the minimum distance from the sun), its value being 1321 kWh/m^2 in the summer solstice, 1366 in the equinoxes, and 1412 in the winter solstice. Using these values and simply geometrical analysis, we could calculate the radiation for day for m^2. However, these values are far from being realistic and important corrections come from the weather conditions, the cloudiness, and the presence of diffused light. To obtain these data, we use PVGIS [4] as an easy instrument, available to everybody, for calculating the energy harvesting for PV systems at different latitude and longitude.

In Fig. 3.2, the energy harvesting in kWh/kWp is given for 25 locations in the Northern hemisphere for south-oriented fixed PV modules with optimal tilt. This plot shows that the weather conditions are very important; as an example, the first two points on the left are for Singapore and Mogadiscio which have the same latitude but a very different energy yield.

The irregular behavior is quite evident. As an example, the yearly energy yield goes from a value around 1800 kWh/y (Northeast Africa and Middle East) to 1000−1200 in South East Asia where monsoon is important, even if, on average, the decrease of energy harvesting with the latitude increase is quite evident.

The energy harvesting can be strongly improved using tracking system, and we are going to analyze one axis tracking and in particular HAT and VAT.

Floating PV Plants. https://doi.org/10.1016/B978-0-12-817061-8.00003-8

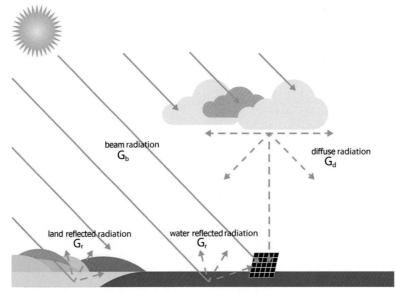

FIG. 3.1 Solar radiation components striking on the ground and on the photovoltaic module [3].

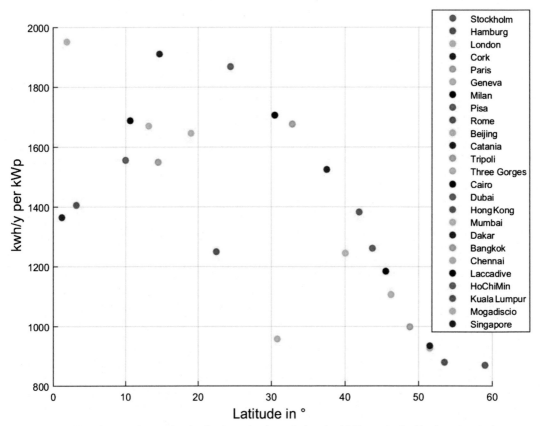

FIG. 3.2 Yearly energy harvesting for fixed south-oriented plant for 25 Towns in the Northern hemisphere.

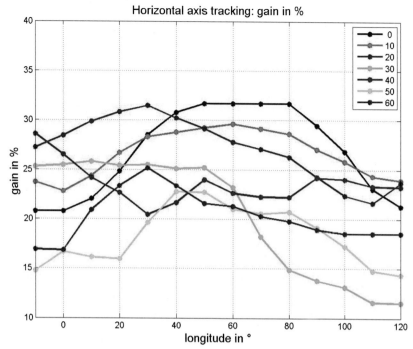

FIG. 3.3 Gain in percentage for PV plants with horizontal axis tracking for latitudes ranging from 0° to 60°.

In Fig. 3.3, the HAT gain with respect to the fixed south-oriented systems with optimal tilt is given versus longitude for different latitudes in the Northern hemisphere. We take into consideration only HAT with angle zero because this solution is particularly suitable for FPV systems, whereas inclined axis implies more expensive solutions and problems of structure.

It is quite evident that there is a remarkable gain for low latitudes which can be more than 30%. However, this advantage is strongly reduced depending on weather conditions and in particular for eastern regions where monsoons are active.

The same results are shown in Fig. 3.4 for VAT systems. Due to the characteristics of the FPV plants which are much more compact than land-based plants, we do not take into account optimal VAT tilt which should imply important shadowing effects, but we consider 20° tilt for VAT systems in the latitude range 0−20°, 30° tilt for latitudes between 30° and 40°, and 40° tilt for higher latitudes.

Results are shown in Fig. 3.4. The importance of the latitude is quite evident: it favors high values and discourages using this solution below 30° of latitude.

Actually, the highest values of the VAT system are reached for Sweden, where high latitude and clear sky combine to give a 35% gain in the energy harvesting with respect to a fixed south-oriented PV system.

3. THE ALBEDO COMPONENT

Recently bifacial modules have become quite popular, and the advantages due to the capture of radiation on the rear part of the modules have been studied. Bifacial solar modules can absorb and convert solar irradiance to current on both their front side and rear side. Several elements affect the bifacial yield, especially the ground albedo around the PV plant.

Several advantages come from bifacial panels [5]:
1. A higher efficiency and a reduction of the gray energy and costs
2. The better exploitation of albedo effects
3. The possibility to better exploit reflector effects [6]
4. The possibility to use it on floating plant

This revives the studies on the albedo effect whose main interest was until now related mainly to the greenhouse effect and to the earth energy balance.

The measure of albedo is given as the percentage of solar radiation reflected, and albedo assumes very

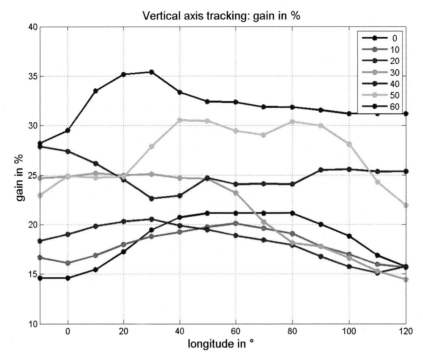

FIG. 3.4 Gain in percentage for the vertical axis tracking.

different values depending on the surface structure. Table 3.1 summarizes the main albedo values for different earth surfaces. Values are taken from many references and are merely indicative, but some comments are in order:

- The albedo value of snow, ice, and desert sand, together with cloud reflection, explains the high value of the earth albedo (30% compared with the moon 12%).
- The ocean albedo is quite low and the upper value is mainly related to whitecaps.
- The albedo on small, calm water surfaces is strongly dependent on the radiation angle (Fresnel law). When the incident angle is low (high radiation

TABLE 3.1
Albedo Values [7].

Albedo Value in %	Min	Max	Albedo Value in %	Minimum	Maximum
New asphalt	4%	6%	Desert sand	40%	50%
Old asphalt	10%	15%	Fresh snow	80%	90%
Macadam	18%		Old snow	45%	70%
Clean cement	55%		Ice	20%	45%
Coniferous forest	8%	15%			
Deciduous trees	15%	18%	Free water surface	5%	
Dry grass	28%	32%	Ocean	5%	10%
Lawns	18%	23%	**Moon global**	**12%**	
tTundra	20%		**Earth global**	**30%**	

intensity), the water reflectivity is very low, around 2%, but for high incident angle it rises and going up to 90 degrees it approaches 1. This condition never happens on the ocean where the waves inhibit the very low angle glint but can happen on a flat calm basin surface.

A detailed analysis of albedo effects on the PV plants has been conducted by several authors [8,9].

Dependence on wind has been studied by several authors (see Refs. [10,11]) and more recently a systematic study has been performed in the United States [12] for Ocean. The water albedo has been modeled as a weighted sum of three components: sun glint, whitecaps, and water-leaving reflectance.

All these complex analysis are in agreement and converge toward a final conclusion: albedo on water has a complex dependence on light wave length, on wind intensity, and on characteristics of water surfaces (waves and sun glint), but it is always rather limited, and for azimuth angles larger than 45° it is on average less than the 5%.

Due to the importance of ocean surfaces on the earth planet, most analyses in the scientific literature are concentrated on the ocean albedo measurements [13]. Fig. 3.5 shows result taken from PV Lighthouse website [14]. It is quite evident that sea albedo is quite depressed except for an elevation angle below 30 degrees (see, for example, [12]). We base our analysis on

recent satellite data taken in the northern hemisphere with different methods [15]. Results of analysis in Ref. [15] come from a long series of data which have been analyzed with two different models COVE and ARPAGE. Without entering into the detail of these models, we only remark that they match quite well and that they can be nicely 11 fitted by a simple periodic plot. The albedo during winter months is more intense due to the fact that in this period the ocean is frequently rough and whitecaps increase the albedo (Fig. 3.6).

Simple Calculation of Albedo on Calm Water Surface

At variance with the cases discussed above the albedo on a calm water surface can be calculated with simple models. The Fresnel laws applied at the discontinuity between air and water (refraction index $n_r = 1.33$) give the albedo coefficients for the solar radiation along the day for a given location, and the albedo due to direct light beam can be calculated with reasonable precision. In Fig. 3.7, three quantities are given for Hamburg, latitude 53.57°:

1. Total average radiation in W/day/m² (divided by 10 for scale reason) for the months of December, March, and January
2. Reflection due to the Fresnel law in percentage for the corresponding solar altitudes
3. Albedo (due to the direct line) in W/m² for the same months

As evident, the very high percentage of reflected light when the sun is low on the horizon gives a limited contribution to the albedo and in average the full contribution due to direct light is essentially the same and is stable around 5%. See Table 3.2.

This result is only for direct solar radiation and has to be corrected by the presence of diffuse radiation.

In order to estimate the albedo of diffused radiation, we perform a first calculation assuming that diffuse light is completely isotropic.

This is not exactly correct, but, in this case, the total result would be given by the integral of the total Fresnel reflectivity $R_{tot}(\theta)$ as function of the elevation angle on all the isotropic directions. The numerical value of this integral is 10.2%. However, this result is misleading since, as proved by measurements [16], the diffuse radiation is more intense in the sunlight direction and for most horizontal radiation its value is reduced by about 50%. Following this line of analysis, we find that albedo strongly increases for low angles, but the average value is 5.5% in agreement with what found for ocean albedo on cloudy days.

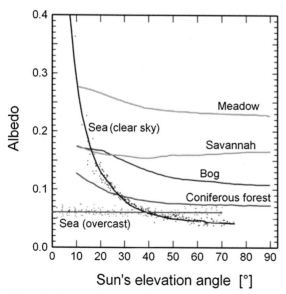

FIG. 3.5 Albedo of Ocean and land surfaces versus incident angle [14].

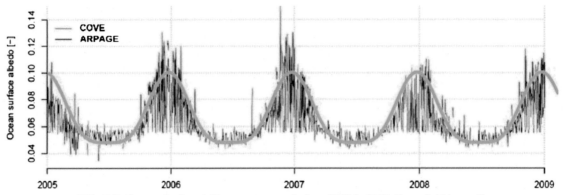

FIG. 3.6 Ocean albedo satellites measurements from 2005 to 2009. Green light is our fit.

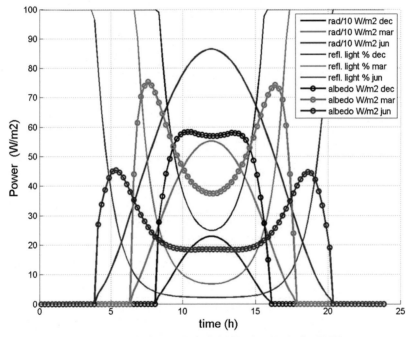

FIG. 3.7 Albedo due to direct light Hamburg latitude, 53.57°.

With this result, we can now calculate the albedo contribution for different locations and we will choose two other typical locations:

- Catania: latitude 34.7°, Mediterranean area, diffuse light below 40%.
- Singapore: latitude 1.2°, Equatorial climate, diffuse light around 50%.

  The calculation has been done in two steps:
- Calculation of the albedo for the direct radiation. The direct radiation as well as diffused radiation data has been taken from PVGIS database.

- Calculation of albedo for diffused radiation as shown in Fig. 3.8. The two calculations have been done for the average of each month and then averaged together.
- Results are shown in Fig. 3.8 for Singapore and in Fig. 3.9 for Catania.

The results should be compared with what has been found by direct measurements. We have results for Singapore in Fig. 3.10 [17]. These data match our calculations and confirms that albedo on water is much less than albedo "onshore" for typical land-based plants.

TABLE 3.2
Solar Energy and Albedo Energy in Hamburg.

Month	Solar Radiation, kWh/day/m²	Albedo Radiation, kWh/day/m²	Rate, %
December	2.2	0.3	13.4%
March	4.9	0.31	6.3%
June	7.9	0.33	4.1%
Yearly average	6.4	0.32	5.8%

These results are rather conclusive: albedo contribution to solar radiation on water is intrinsically limited so that the use of bifacial PV modules does not give great advantages in the FPV plants. However, FPV offers other opportunities and the coupling of FPV with bifacial and reflectors will be discussed in Chapter 7 for HAT systems.

4. GEOGRAPHIC POTENTIAL AND FRESHWATER SURFACES AVAILABILITY

As widely discussed in Ref. [3], one of the main problems of a PV system is the land occupancy related to the efficiency of the solar energy conversion in electricity and to the PV system characteristics [18].

The quantities useful for defining the PV geographic potential, PVGP, are the annual irradiation in kWh/m², on a horizontal surface multiplied by the area suitable for installing the PV, A_s.

The ratio between A_s and the total land area under consideration, A_L is very important, so we introduce it explicitly as follows: $\alpha_{SL} = A_s/A_L$. The same definition can be extended to the surface covered by water in a given territory, A_w, that can be used for FPV plants, $\alpha_{SW} = A_s/A_W$.

The concept of available area is the first input of the more useful technical PV potential that can be calculated in terms of capacity (installed power) or generated electricity (energy):

- PVPP$_d$ (PV power potential density in kWp/m²)
- PVEP$_d$ (PV energy potential density in kWh/m²/year)

These are defined as the PV power per unit of land (water) area and the PV energy per unit of land (water) area, respectively. Of course, the PVPP (PV power potential) and the PVEP (PV energy potential) are given, respectively, by PVPP$_d$·A_s and by PVEP$_d$·A_s.

The latter is the actual useable solar power or yearly energy yield, once it has been transformed into electricity by a PV system.

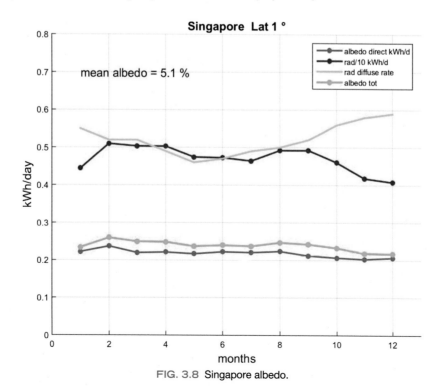

FIG. 3.8 Singapore albedo.

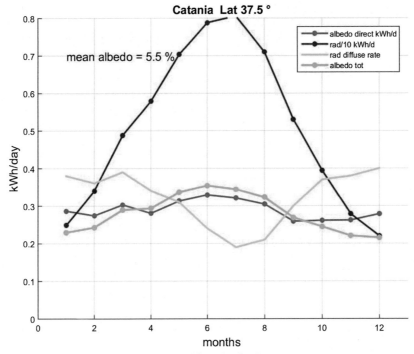

FIG. 3.9 Catania albedo.

The PVPP$_d$ depends on many parameters, but in the following, we will assume that standard PV modules of 1×2 m surface are used, with a given power per panel. This power until now has ranged between 300 and 400 Watt, but it is slowly increasing; 500 W modules will be on the market by 2020.

Finally, we must consider the packing factor PF taking into account the structure of the plant, the modules tilt angle, and the necessity to avoid shadows between modules, as well as the necessity to have space for managing purposes: $PF = A_{PV}/A_{GEN}$, where A_{PV} is the modules surface and the projection on the horizontal plane of the PV modules area and A_{GEN} is the area covered by the PV generator system which, in a first approximation, can be considered equal to A_s.

The value of PVPP$_d$ strongly depends on the technology used and on the choice of system FIX, VAT, or HAT.

For land-based PV plants, typical values are PVPP$_d$ = 700−800 kWp per ha. This value rises to more than 1 MWp for floating plant (where saving raft surface is a must) and goes up to 1.8 MWp per ha for very compact floating structures. Of course, HAT and VAT systems require a wider surface in order to optimize energy harvesting and reduce the shadow effect. These values will be discussed in the chapter of structures and of tracking.

How large are the freshwater surfaces available worldwide?

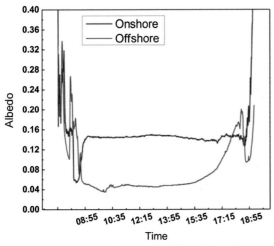

FIG. 3.10 Albedo in Singapore SERIS test bed. Albedo on land is in red (March 15, 2017).

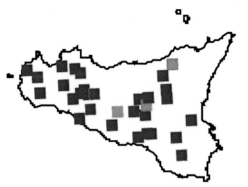

FIG. 3.11 Sicily artificial (red) and natural (green) basins (surface > 20 ha).

We have collected in Table 3.3 the main world areas, and even if a myriad of small irrigation basins and water reservoirs are neglected, the numbers give an idea of the order of magnitude of the available water surfaces [19].

The second column gives the geographic PV potential assuming that an average value of $\alpha_{SW} = 1\%$ can be used for installing floating plants.

These values are surprisingly high and a more detailed analysis can show that the quoted values are a lower limit to the real potential.

Sicily Basins

As an example, let us analyze basins in Italy and specifically in one of the driest Italian regions, Sicily.

Sicily, one of the most arid regions in Italy, has a lot of artificial (red dots) and natural (green dots) basins. The map in Fig. 3.11 gives the locations of basins of more than 20 ha on Sicily and highlights the fact that even in the south the number of artificial and natural lakes is very high [20].

A complete list is given below for Sicily in Table 3.4.

In conclusion, the full available water surface in Sicily is about 75 km^2, constituted mainly by small artificial basins which can be partially covered without any problem. So assuming a value of $\alpha_{SW} = \mathbf{10\%}$, this implies a **PVPP** value of more than 1 GWp. However, this is only a part of the full story. We should note that small basins for irrigation and wastewater treatment are not taken into account in these data table, as well as artificial basins created by human activities (industry, quarry basins, irrigation reservoirs, etc.).

Small Basins due to Human Activities

In order to illustrate the problem, we give in Fig. 3.12 a list of 20 basins located to the west of Treviso in an area between Treviso and Castelfranco Veneto (Table 3.5).

The analysis is not exhaustive (several small basins are not highlighted), but this gives an idea of the great availability of water surfaces originated mostly from quarries or irrigation basins.

Most of these basins are abandoned and would require maintenance or in any case structural managing, and we have assumed that a coverage of 50% of their surface would be a convenient choice. In this case, 124 ha of 247.7 ha should be available and would allow the installation of 167 MWp without any land occupancy, with environmental benefits and with a production of more than 200 GWh per year.

This analysis has been done for several small, medium size Italian towns (Pisa, Pavia, Brescia, etc.) with analogous results. For any medium town in Italy (less than 200,000 inhabitants), the available basins, within 20 km radius from the city center, allow the installation of at least 100 MWp with relevant environmental benefits. The same situation holds true going up with latitude and looking at central and northern Europe.

We want to stress further that none of these basins is included in previous statistics which concern only basins exceeding 20 ha.

TABLE 3.3				
Worldwide Geographic Power and Energy Potential.				
	Water Surface	α_{SW}	**PVPP**	**PVEP**
	km^2	1%	GW	TW
Tropical	1.254.831	12.548	1.526	1.851
Temperate	1.506.256	15.063	1.832	2.064
Cold zone	1.789.819	17.898	2.176	2.155
Total	**4.550.906**	**45.509**	**5.534**	**6.069**

PVEP, PV energy potential; *PVPP*, PV power potential.

TABLE 3.4
Sicily Basins Following LIMNO.

Sicily Basins	Surface, km²	Volume, 10⁶ m³	Sicily Basins	Surface, km²	Volume, 10⁶ m³
Ancipa	1.4	28.1	Paceco	1.2	6.7
Arancio	3.7	34.8	Pergusa (naturale)	1.8	5.8
Biviere di Cesarò (naturale)	0.2	0.1	Piana degli Albanesi	3.3	29.8
Castello	1.6	21	Piana del Leone	0.6	4.2
Cimia	0.9	10	Poma	3.6	72.3
Comunelli	0.9	7.8	Pozzillo	7.8	127.4
Dirillo	1.1	20.1	Prizzi	0.9	9.3
Disueri	1.5	14	Rubino	1.4	11.5
Fanaco	1.4	20.7	San Giovanni	2	16
Gammauta	0.2	0.8	Santa Rosalia	1.3	20
Garcia	5.1	80	Scanzano	1.5	17.3
Gorgo	0.5	3.1	Sciaguana	1	11.4
Lentini	10	127	Soprano (naturale)	0.2	0.2
Nicoletti	1.6	20.2	Trinita'	1.9	18
Ogliastro	14	110	Villarosa	1.3	15.4
Olivo	1.1	15	**Total**	**74.9**	**877.8**

FIG. 3.12 Map of 20 basins to the west of Treviso.

TABLE 3.5
Treviso West: Water Reservoir.

Basin	ha	$\alpha_{sw} = 50\%$ PVPP, MWp	Energy PVEP, GWh	Basin	ha	$\alpha_{sw} = 50\%$ PVPP, MWp	Energy PVEP, GWh
TV1	8	5.40	6.8	TV11	7.4	5.00	6.3
TV2	14	9.46	11.9	TV12	2.6	1.76	2.2
TV3	24	16.21	20.5	TV13	14.5	9.80	12.4
TV4	19	12.84	16.2	TV14	9	6.08	7.7
TV5	9.4	6.35	8.0	TV15	6.5	4.39	5.5
TV6	4.2	2.84	3.6	TV16	14	9.46	11.9
TV7	12.7	8.58	10.8	TV17	24.5	16.55	20.9
TV8	3.5	2.36	3.0	TV18	7	4.73	6.0
TV9	9.4	6.35	8.0	TV19	37.7	25.47	32.2
TV10	15.5	10.47	13.2	TV20	4.8	3.24	4.1
				Total	**247.7**	**167**	**211**

PVEP, PV energy potential; *PVPP*, PV power potential.

Pakistan Channels

Other water surfaces can be used for installing FPV. Channels are a typical structure which can benefit from a partial coverage with FPV plants. Pakistan is crossed by many long channels going from the south to the north. Statistics shown in Fig. 3.13 puts in evidence that several GWp can be installed on channels in Pakistan reducing water evaporation and mitigating algal bloom problem. See Ref. [21].

Wastewater Basins

An analogous problem arises when considering wastewater treatment plants. They are absolutely necessary

and are frequently located not far from populous cities. The main problems are related to the reuse of water for agricultural purpose which in arid climates is strongly reduced by evaporation and to the unpleasant odors which can arise during the bacteria digestion of the polluted waters.

We have studied in Ref. [3] several Australia wastewater basins as useful examples. Here we extend this analysis to Qatar in order to show that even in very desertic areas, where there are human settlements, there are also wastewater basins which can be usefully exploited for energy production with the advantage of saving a lot of precious water resources. As an example,

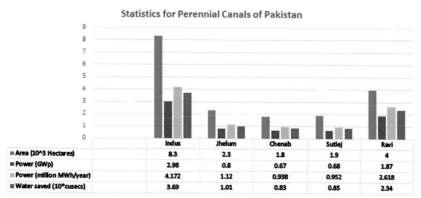

FIG. 3.13 Water surface of channels in Pakistan analyzed for the main five districts.

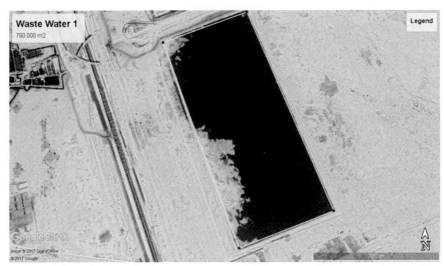

FIG. 3.14 Basin of waste water near Doha: 25° 11′ 54″ North, 51° 19′ 32″ East.

an FPV plant covering the large basin near Doha, shown in Fig. 3.14, will give rise to a power of about 100 MWp. Such systems have two advantages:

- The cost which is comparable with that of a standard land-based plant but with a cheaper maintenance and a better energy harvesting.
- The strong reduction of the evaporation rate which allows the saving of 2.5 million of m³ of water per year for each hectare covered by an FPV plant.

Another larger basin (100 km from Doha) is shown in Fig. 3.15.

In this case the available surface is 480 ha and the PVPP exceeds 500 MWp.

But several other wastewater basins exist in Qatar, and by exploiting part of the existing surfaces, more

than 1 GW of PV modules could be installed, saving 20 million of cubic meters of water every year, which would become available for agricultural uses.

5. SALTY WATER

As far as seawater systems are concerned there are several possible solutions and we should distinguish between low depth inlets of the sea or lagoons and offshore. The latter aspect will be discussed in Chapter 11.

Coastal Water and Atolls

Only in the Mediterranean area [22] about 6400 km² of coastal lagoons are listed, sometimes abandoned, sometimes used as fisheries, or protected as areas interesting for biodiversity. See Fig. 3.16.

But in other areas of the world, the lagoon extensions are very large. In Emirates and Qatar, the costal structure is a unique immense lagoon whose surface is approximately 50,000 km² with the potential of many tenths of GWp of FPV plants.

More generally speaking, coastal areas comprise 20% of the Earth's surface and yet contain over 50% of the entire human population. By the year 2025, coastal populations are expected to account for 75% of the total world population. Worldwide, over 1200 major estuaries, lagoons, and fiords have been identified covering an area of about 500,000 km² [23].

In other cases, such as atolls, the main problem is related to the limit in waves and to the proximity of electrical connections with the grid network. In these

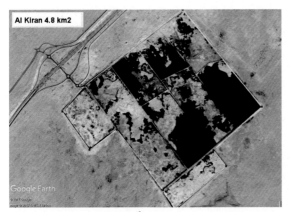

FIG. 3.15 Basin of 4.8 Km²: Alkiran 24° 58′ 47″ North 51° 02′ 38″ East.

FIG. 3.16 Coastal lagoon: Cadiz Bay, Spain.

TABLE 3.6
List of the Largest Atolls in the World [24].

Largest Atolls in the world	Position	Area (km^2)
Great Chagos Bank	6.17°S 72.00°E	12,642
Reed Bank	11.45°N 116.90°E	8866
Macclesfield Bank	16.00°N 114.50°E	6448
North Bank Saya de Malha Bank	9.07°S 60.20°E	5800
Rosalind Bank	16.43°N 80.52°W	4500
Boduthiladhunmathi	6.73°N 73.04°E	3850
Chesterfield Islands	19.35°S 158.66°E	3500
Huvadhu Atoll	0.50°N 73.30°E	3152
Truk Lagoon	7.42°N 151.78°E	3152
Sabalana Islands	6.75°S 118.83°E	2694
Lihou Reef	17.42°S 151.67°E	2529
Bassas de Pedro	13.08°N 72.42°E	2474
Ardasier Bank	7.71°N 114.25°E	2347
Kwajalein	9.19°N 167.47°E	2304
Diamond Islets Bank	17.42°S 150.96°E	2282
Namonuito Atoll	8.67°N 150.00°E	2267
Ari Atoll	3.86°N 72.83°E	2252
Maro Reef	25.42°N 170.59°W	1934
Rangiroa	15.13°S 147.65°W	1762
Kolhumadulhu Atoll	2.37°N 73.12°E	1617
Kaafu Atoll (North Malé Atoll)	4.42°N 73.50°E	1565
Ontong Java	5.27°S 159.35°E	1500
Total		**76,963**

cases, the size of the floating plant can only be based on ad hoc projects.

In Table 3.6, we give the largest atolls in the world; however, thousands of small atolls are missed in this list and could be very interesting as a solution for the local inhabitants' energy needs.

As for freshwater basins, this is only the tip of the iceberg, and many solutions are possible adapting the very flexible FPV structure to the local context and to the needs of the inhabitants.

BIBLIOGRAPHY

[1] J.A. Duffie, W.A. Beckman, Solar Engineering of Thermal Processes, John Wiley & Sons, New York, USA, 1980.

[2] G.M. Masters, Renewable and Efficient Electric Power Systems, John Wiley & Sons, New York USA, 2013.

[3] M. Rosa-Clot, G.M. Tina, Submerged and Floating Photovoltaic Systems, Modelling, Design, Case Studies, Elsevier, Academic Press, London, 2017.

[4] Photovoltaic Geographical Information System, PVGIS [Online]. Available: https://re.jrc.ec.europa.eu/pvgis/.

[5] P. Rothhardt, S. Meier, K. Jiang, A. Wolf, D. Biro, 19.9% efficient bifacial solar cell produced by co-diffusion, in: 29th European Photovoltaic Solar Energy Conference and Exhibition, Amsterdam, 2014.

[6] J. Castillo-Aguillea, P. Hauser, Multi-variable bifacial photovoltaic module test results and best-fit annual bifacial, IEEE Access (2016).

[7] Y. Kotak, M.S. Gul and T. Muneer, Investigating the impact of ground Albedo on the performance of PV systems.

[8] R.W. Andrews, J.M. Pearce, The effect of spectral albedo on amorphous silicon and crystalline silicon solar photovoltaic device performance, Solar Energy 91 (2013) 233−241.

[9] M.P. Brennan, A.L. Abramase, R.W. Andrews, J.M. Pearce, Effects of spectral albedo on photovoltaic devices, Solar Energy Materials and Solar Cells 124 (2014) 111−116.

[10] C.A. Varotsos, I.N. Melnikova, A.P. Cracknell, C. Tzanis, A.V. Vasilyev, New spectral functions of the near-ground albedo derived from aircraft diffraction spectrometer observations, Atmospheric Chemistry and Physics 14 (2014) 6953−6965.

[11] K. Katsaros, L. McMurdie, J. Devault, Albedo of a water surface, spectral variation, effects of atmospheric transmittance, sun angle and wind speed, Journal of Geosphysical Research 90 (1985) 7313−7321.

[12] Y. Feng, Q. Liu, Y. Qu, S. Liang, Estimation of the ocean water albedo from remote sensing and meteorological reanalysis data, IEEE Transaction on Geoscience and Remotesensing 54 (2) (2016) 850−868.

[13] Y. Qu, S. Yang, Q. Liu, X. Li, Mapping surface boradband Albedo from satellite observation: a review of literature on algorithms and products, Remote Sensing 7 (2015) 990−1020.

[14] https://www.pvlighthouse.com.au/,» PV lighthouse consulting, [Online]. Available: https://www2.pvlighthouse.com.au/resources/courses/altermatt/The%20Solar%20Spectrum/Albedo.aspx.

[15] R. Seferian, S. Baek, O. Boucher, J. Dufrene, B. Decharme, D. SainMartin, R. Roeherig, An interactive ocean surface albedo scheme (OSAv1.0): formulation, Geoscientific Model Development 11 (2016) 321−338.

[16] E. Andersen, K. Nielsen, J. Dragsted, S. Furbo, Measurements of the angular distribution of diffuse irradiance, Energy Procedia 70 (2015) 729−736.

[17] T. Reindl, At the Hearth of Floating Solar: Singapore, February 2018, pp. 18−23. www.pv-tech.org.

[18] N. Martín-Chivelet, Photovoltaic potential and land-use estimation methodology, Energy 94 (2016) 233−242.

[19] C. I. A. US, The World Facebook, 2017. [Online]. Available: https://www.cia.gov/library/publications/the-world-factbook/. [Consultato il giorno 2017].

[20] LIMNO, database della qualità dei laghi italiani, [Online]. Available: http://www.ise.cnr.it/limno/limno.htm. [Consultato il giorno 18 june 2019].

[21] A. Naveed, H. Qamar, M. Rosa-Clot, A Case for Using Solar Floating PV to Meet Future Peak Energy Demand in Pakistan, 2019.

[22] Mediterranean, Genearl, Fisheries e Commission, Mediterranean Costal Lagoons, FAO, 2015.

[23] World Ocean Network. [Online]. Available: https://www.worldoceannetwork.org/won-part-6/carem-wod-2014-4/thematic-resources-coastal-management/facts-figures-coastal-management/.

[24] Wikipedia. [Online]. Available: https://en.wikipedia.org/wiki/Atoll.

CHAPTER 4

Floating PV Structures

RANIERO CAZZANIGA, CTO, R&D

1. INTRODUCTION

Floating photovoltaic (FPV) is no more an eccentric idea of some scholars but an actual commercial product that has already seen real plants built all over the world. However, so far, there are a number of very different floating technologies that are improving, and studies are needed to find a consolidated standard. The authors are working in this field since 2008, and they are still working trying to set these standards.

From the very beginning, it was evident that there were three distinct options:

Class 1: High-density polyethylene (HDPE) pipes plus steel or aluminum components for building rafts of large dimensions

Class 2: Full HDPE rafts of small dimension, typically mono-modules connected together by suitable hooks.

Class 3: Floating pontoon structures, connected together and able to support photovoltaic (PV) modules.

A typical structure of the first class that has been used by Terra Moretti in several projects with the collaboration of Koiné Multimedia (see Figs. 4.1 and 4.2):

The second one was first proposed by Ciel & Terre, and it is the more commonly adopted up to now together with a large set of competitors and imitators (Fig. 4.3).

The name of this technology is Hydrelio with manufacturing lines in seven countries on four different continents (Figs. 4.4 and 4.5).

Class 3 has a typical example in one of the first plants built in Italy in Bubano in 2009 by NRG Energia (Figs. 4.6 and 4.7).

These three approaches have had a lot of imitators with sometimes, mixed solution, and it is not easy to compare all these. Actually the main elements for judging the merits of the different solutions are:

- Robustness
- Simplicity in assembling, launching, and mooring
- Possibility to adapt the system to local conditions (panel tilt, cooling, tracking, etc.)
- Minimum environment impact and minimum water contamination
- Cost

Due to the very quick growth of the sector, it is impossible to list all the companies working in this field, but it is, however, useful to list the more prominent names highlighting advantages and limits with reference to the items outlined above.

After this list, a detailed analysis of new Class 1 solutions will be given.

2. SOME OF THE MOST INTERESTING FPV SOLUTIONS

In the following, we will list several solutions proposed by different companies, and we will group them using the classification quoted above.

2.1 Class 1

The Suvereto project, 200kWp (Fig. 4.1), is grid connected since 2011 and in almost 9 years of service demonstrated itself very robust. A few years later a monoaxial tracking system has been added (Koiné design), improving the harvesting. Since the structure is still in pristine conditions, it could reach a power of 450 kW simply using the new panels that are now available.

Benefiting from the experience made with the Colignola plant (equipped with cooling system, tracking system, and reflectors) in 2012, a new project was implemented in Korea, probably the first Korean plant of that kind at the time.

In this installation each raft supports six panels with a 35 degrees tilt, and it has flat reflectors on the rear. This structure has been adopted with minor modifications in several proposals using different materials. See Ref. [1] where extruded beams in plastic materials are suggested. Efforts have been made to build rafts supporting a few panels (2−4) with several materials:

FIG. 4.1 Suvereto plant (Terra Moretti and then Koiné).

FIG. 4.2 La Badiola plant under construction (Terra Moretti).

FIG. 4.3 Ciel & Terre proposal.

HDPE, PVC, or, as in the Korean project case, a mix of HDPE and galvanized steel [2] (Figs. 4.9 and 4.10).

The system is very robust, and each raft supports six panels. Assembly and launching are quite simple, but the system is rather expensive if we consider today's targets.

The system is perfectly walkable and has a large buoyancy and last, but not least, the contact surface between plastic pipes and water is roughly only 20% of the plant surface.

Four examples of the Class 1 system are following.

Terra Moretti (Fig. 4.11)

K-Water solatus

K-Water solutions in part imitate the Koiné design, but this company has also studied several other solutions. Along with Class 1 designs we also find Class 2 and Class 3 solutions (Fig. 4.12).

Fig. 4.13 shows details of the 3-MWp PV plant assembled on Jipyeong Reservoir, Sangju City (Korea) by the Korean company LG CNS [3]. This technology looks quite similar to solutions proposed by K-Water.

Sunfloat

This is a very recent proposal suggested for Dutch large basins and specifically applicable to bifacial PV modules [4].

The Sunfloat solution resembles the FPV structure previously shown in Fig. 4.8 and probably has the same high cost limitations even if robustness is a strong feature (we do not have details of this prototype) (Fig. 4.14).

4C Solar

4C Solar develops the concept of floating pipes further by superposing a steel or aluminum structure [5]. A near offshore system has also been studied (Fig. 4.15).

2.2 Class 2

The solution first proposed by Ciel & Terre and then, subsequently, imitated by many competitors has the largest number of installations with remarkable market success: in 2016—18, several hundreds of MWp of this kind of systems were installed worldwide, mainly in China and South East Asia.

The limitation of this solution is linked to the same aspect that made it so appreciated: the lightness of the

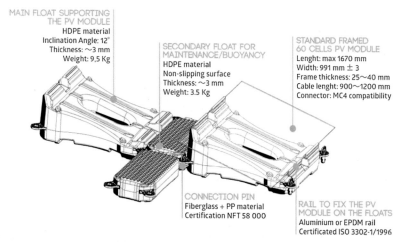

MAIN FLOAT SUPPORTING
THE PV MODULE
HDPE material
Inclination Angle: 12°
Thickness: ~3 mm
Weight: 9,5 Kg

SECONDARY FLOAT FOR
MAINTENANCE/BUOYANCY
HDPE material
Non-slipping surface
Thickness: ~3 mm
Weight: 3.5 Kg

STANDARD FRAMED
60 CELLS PV MODULE
Lenght: max 1670 mm
Width: 991 mm ± 3
Frame thickness: 25~40 mm
Cable lenght: 900~1200 mm
Connector: MC4 compatibility

CONNECTION PIN
Fiberglass + PP material
Certification NFT 58 000

RAIL TO FIX THE PV
MODULE ON THE FLOATS
Aluminium or EPDM rail
Certificated ISO 3302-1/1996

FIG. 4.4 Ciel & Terre: Hydrelio scheme.

FIG. 4.5 Hydrelio locking bolts.

FIG. 4.6 SRG proposal (Bubano plant 2010).

rafts exposes the structure to the stresses of wind and wave loads, and a storm can damage the rafts or the full structure.

The floaters are very thin, and water-plastic contact is very large, at least 50% of the plant surface, with possible problems in terms of long-term plastic defoliation.

All these kinds of systems are molded, and the molds are very expensive so they cannot be customized and a tracking system is not applicable to the structure. The mooring system is extremely complex: many cables are necessary to secure the platform and for a large surface it is not sufficient to anchor the system on the perimeter, and it is necessary to use most of the cables underneath the structure.

A single floater cannot support a person, and then walkability and maintenance suffer some limitations. The combination of all these elements give rise to events like the one shown in Fig. 4.16.

The main advantage of this solution lies in the cost which is quite limited, and can be assumed to be 25 c$ per Watt installed. This value is just an average of several pieces of information garnered from the market and can vary depending on the market conditions and in particular on the oil price; since the total system is made from polyethylene the price is linearly and directly linked to the price of a barrel of oil at any given time.

We quote here four of the many companies working with this technology: Sunseap, Isifloating, Sumimoto, ATS/ASB.

Sunseap solution

This solution (Ciel & Terre collaboration) has been tested at the SERIS test bed [6]. In Fig. 4.17, a 5-MWp plant is shown with the necessary perimeter rafts added for the protection of the plant edges.

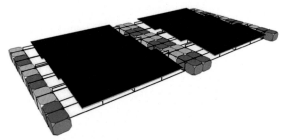

FIG. 4.7 Scheme of the engineering of Bubano plant.

ATS/HDB

The Housing and Development Board (HDB) is also exploring floating solar panels in open sea [8] (Fig. 4.21).

2.3 Class 3

A solution to the safety problems associated with Class 2 FPV plants is provided by Class 3 plants. In this case, the innovation consists in building a rather large platform that is quite rigid and able to support many PV modules (10–100). The platform is sup-

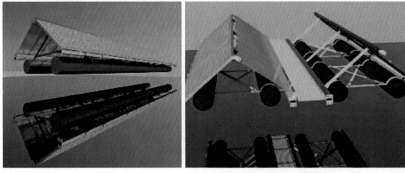

FIG. 4.8 Korea project 2012 (Koiné design).

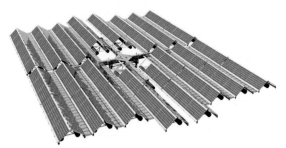

FIG. 4.9 CAD of the Korea pilot installation, 2012.

IsiFloating

The IsiFloating concept is quite similar to the Hydrelio solution even if it seems a more robust and compact system (Figs. 4.18 and 4.19).

Sumitomo Mitsui Class 2

The photo in Fig. 4.20 exhibits the walkability of the full platform that has to be facilitated by the addition of extra mono-module rafts [7].

FIG. 4.10 The Korean platform in winter.

ported by floating elements which have different shape usually originating from standard marine pontoon and oil platform technologies.

FIG. 4.11 Terra Moretti structure of rafts during assembly.

This is the sector offering the largest variety of designs, and we quote here seven of the many companies suggesting this solution (Table 4.1).

All these systems have as their objective the building of large size platforms where the ability to walk on the structure and the structural buoyancy is guaranteed. This result is achieved, but the costs are on average much higher than for Class 2 solutions and the construction, in several cases, is rather complex.

Adtechindia
This company suggests the use of Ferrocement technology. Fig. 4.22 shows an aerial photo of a 500-kWp platform built in India [9].

NRG-Energia
This is one of the first concepts in Class-3 and it has been initially developed in Bubano (Italy) in 2009. In Fig. 4.23, an assembly of two units of 40 PV modules is shown [10].

Takiron Energy
Fig. 4.24 shows the assembling of an FPV platform using floats manufactured by Takiron Engineering Co. Ltd of Kita-ku, Osaka City [11].

Swimsol
The Swimsol solution has been proposed for salty water and for offshore solutions. There are no details on the floating structure which appears to be a large rigid structure that has the ability to support 24-kWp PV modules. Four of these systems were installed in the Baa Atoll (Maldives) (Fig. 4.25) [12].

Solaris Float
Fig. 4.26 shows a rather original structure where small floating elements are rigidly connected and generate a full circular platform. This system facilitates the fitting of a double axis tracking system [13].

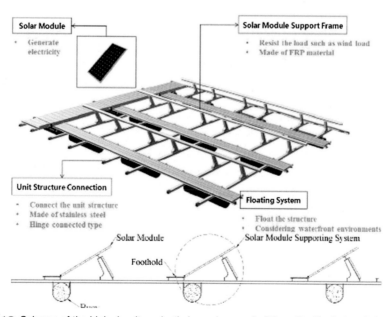

FIG. 4.12 Scheme of the high-density polyethylene pipes and of the raft with photovoltaic modules.

FIG. 4.13 LG CNS staff inspecting the floating photovoltaic modules.

FIG. 4.14 Sunfloat prototype.

FIG. 4.15 4C Solar structure.

Nemo-Eng
NEMO Eng proposes to use a metal-framed floating body that provides stability and prevents damages from UV rays, temperature changes, and external shocks (Fig. 4.27) [14].

Moss Maritime
Moss Maritime builds on more than 20 years of experience in designing and engineering oil and gas floating

FIG. 4.16 Problems on a Class 2 plant.

FIG. 4.17 Sunseap solution: 5-MWp plant.

facilities. The rendering in Fig. 4.28 shows a possibility of extending their technique to the FPV sector [15].

3. UPSOLAR-KOINÉ PROJECT (SINGAPORE)
The main problem of Class 1 solutions lies in the cost and in rafts assembling difficulties. In order to overcome these problems, a new raft concept has been developed and built [16] belonging to Class 1 [17].

Its concept is shown in Fig. 4.29: the raft is about 12 m long and floats on four HDPE pipes. The structure of galvanized steel beams can be easily assembled in a suitable area near the basin without welding activities.

The raft can accomodate two rows of modules in landscape position. The total number of modules ranges between 16 and 24 depending on their tilt. The use of lighter aluminum beams and of smaller HDPE pipes could further improve the cost of the project.

The structure facilitates the wiring of electric cables as well as of other components that can be placed onboard.

FIG. 4.18 IsiFloating Monopanel raft structure.

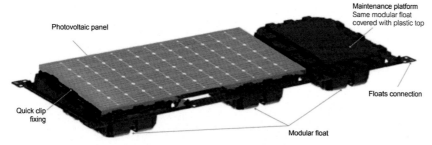

FIG. 4.19 Isifloating: raft design.

FIG. 4.20 Sumitomo full structure.

FIG. 4.21 The floating solar system at Tengeh reservoir.

This solution has been installed at the SERIS test bed in the Tengeh reservoir (Fig. 4.30).

This structure is widely customizable to be adapted to different solutions: Fig. 4.31 shows an example of a 24-panel raft installed in Λ configuration (gable). This provides a reduction of costs per kWp installed and it's particularly suited for tropical regions where the tilt of the modules must be small.

The buoyancy capacity of this raft is large: the weight of raft, modules, and cables included is in the order of 1500 kg, and the volume of the pipes is more than 3 m³. The result is a free buoyancy of almost 2000 Kg that can easily support the weight of two or three people. When more rafts are connected we get a global buoyancy that increases the walkability and improves the safety for people working in plant maintenance.

Fig. 4.32 shows the scheme of the 100-kWp plant built on the Tengeh reservoir with 20 modules per raft.

Furthermore the full plant allows for small movements in the zeta direction, but the platform is quite

TABLE 4.1
List of quoted companies of Class 3.

1	Adtechindia	Large Blocks Supporting 120 PV modules
2	NRG-Energia	Blocks supporting 40 PV modules
3	Takiron Energy	Blocks with 9–12 PV modules
4	Swimsol	Large structures with 48 PV modules
5	Solaris Float	Standard blocks with single panels connected in a large lattice
6	Nemo-Eng	Rigid floating body connected by steel beams
7	Moss Maritime	Platforms of 40 PV modules supported by six large floating blocks

FIG. 4.24 Takiron: the structure.

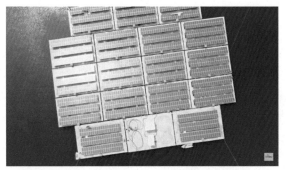

FIG. 4.22 Adtechindia, 500-kWp platform.

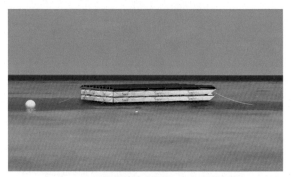

FIG. 4.25 Swimsol floating unit (24 kWp).

FIG. 4.23 NRG-Energia solution: two units of 40 PV modules each.

rigid in the X-Y horizontal plane, and this is useful for moving the platform as a whole, as would be required if a vertical tracking system is installed.

This approach guarantees:

Modularity: The structure is modular. Raft components are pipes and beams of galvanized steel that can be easily assembled in a very short time. Almost all the job and in particular the PV modules are installed and connected while working on land.

Material availability: All the materials can be found locally. HDPE pipes are certified and compatible with drinking water for civil use, and the availability of fabrication workshops is standard in every industrialized and emerging country. All components can easily fit in containers, and 1.5 containers are of sufficient size for a 100-kWp plant and their assembly is quick. At the Tengeh reservoir, 5 days were enough for the completion of emptying containers, mounting the raft, and assembling the PV plant.

Flexibility: The structure can be modified in order to implement cooling and tracking systems.

FIG. 4.26 Solaris Float: two axis tracking system.

FIG. 4.27 Nemo-Eng steel platform with 36 PV modules.

FIG. 4.28 Moss Maritime Ponton rendering.

FIG. 4.29 A single raft in Singapore project.

Environment impact: The contact between pipes and water is only 20% of the full plants surface area.

Robustness: HDPE pipes have a thickness of around 12 mm and are guaranteed 50 years long. Furthermore, the plant is quite heavy and a real situation with wind up to 140 Km/h has been experienced without causing any damage to a sister structure [18].

Safety: Three elements contribute to the safety of the structure: (1) Excellent walkability which allows operators to move safely around the platform carrying out their tasks. This is made possible by the inherent large buoyancy of the structures. See Fig. 4.33. (2) Materials used (HDPE pipes and galvanized steel) that are compatible even with drinking water basins like Tengeh. (3) Electric grounding.

Mooring: The mooring system is very robust and simple and can be installed around the edge of the platform for a plant up to a few MW (Fig. 4.34).

In conclusion, a great advantage of this structure is the robustness and walkability. The main limit is the cost which, on average, remains higher than the Class 2 solutions as it costs around 30 c$ per Watt.

To solve this problem without losing anything in robustness and walkability, a new solution has been proposed and recently patented.

4. GABLE SLENDER SOLUTION

This new design is very different from anything that has been proposed so far.

The only distant kinship can be found in the design proposed, e.g., in Singapore, Fig. 4.29.

For a better understanding of this concept it could be useful a direct comparison with a typical Class 2 product.

The first thing to notice is that a Class 2 has, typically, a 1:2 ratio meaning that for every panel, more than one floater (usually 2) are needed. Because of the huge number of floaters the thickness must be low, typically less

FIG. 4.30 Singapore Tengeh reservoir with the 100-kWp Upsolar-Koiné project in phase of completion.

FIG. 4.33 Walkability of the Koiné-Upsolar at the Tengeh reservoir.

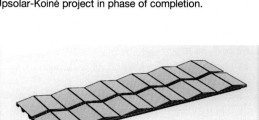

FIG. 4.31 Raft with 24 PV modules tilted to 5 degrees.

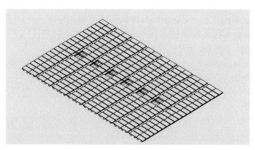

FIG. 4.32 100-kWp plant (Tengeh project).

than 3 mm. If we think about a possible 1-MW plant, this means:

- **Gable slender plant:** 110 strong floaters, 12 mm thick each
- **Typical Class 2 plant:** Up to 5000 floaters, less than 3 mm thick each

All units have to be connected, so in the first case you have to connect 110 strong and heavy rafts (and periodically check all of them later) where in the second case you have to connect thousands of light floaters (and periodically check all of them later). The total number of parts in a mechanical system is always a very important parameter. It is demonstrated that the smaller the

number of parts, the greater the reliability of the structure.

Another important reference, may be the most important, is the mooring design.

Gable slender has a strong metal structure and very strong mooring points. A full-plastic system requires plastic nade mooring points. The single breaking load is inevitably low, then you need to distribute the loads. A very large number of mooring points are then mandatory. The many lines deployed underneath the platform can be placed by scuba divers only and you need scuba divers for all the checking and maintenance operations. The gable slender mooring design provides a mooring on the perimeter, then the checking operations are visually done and all the ordinary maintenance can be done out of the water.

If we get back to the 1 MW, we can compare:

- **Gable slender plant**: 20 mooring lines (only on the perimeter)
- **Typical Class 2 plant**: Up to 200 mooring lines (mainly under the platform)

The gable slender design considers that the best fit on water is a boat. The base unit can be seen as a 12-m-long boat that has room for 24 panels, a few floaters constituted by HDPE pipes and only a few mooring lines on the perimeter, all visible and that can all be inspected.

Fig. 4.35 represents a single "boat," the base unit of the platform.

The typical size of the pipe is 12 m long and 0.5 m in diameter. Thickness can vary between three and four times the size of a class 2 floater. This guarantees a good buoyancy and a very long life cycle. Being the volume of one pipe 2.35 m^3 and the raft weight with PV modules and cables approximately 1300 kg, we have a free buoyancy of 1000 kg per raft. Furthermore, the

FIG. 4.34 Mooring: the eyebolts of the platform.

FIG. 4.35 Gable slender: the raft unit.

contact of the plastic with the water is approximately 20% only of the total raft surface.

All boats are joined at a few points, but the connections allow independent relative rotation. Then the boats are strong, but the platform is flexible (Fig. 4.36).

The system allows the connection of a very large number of boats, both on the long side and on the short side, so as to constitute platforms with power of the order of many MWs (Fig. 4.37).

Another problem solved, thanks to the big size of the single boat, is the "walkability." All FPV plants need to integrate corridors between panel rows because the maintenance team must have some room for walking and then reaching the different areas of the plant. This is a remarkable cost for structures that have no function in the energy harvesting. The big pipes of gable slender, standing between the panels, are easily adaptable for walking at a low cost. Permanent or nonpermanent boards can be put between the rafts where the free space is almost half a meter. Additionally, the design of other

special devices is in an advanced stage of testing to realize automatic operations on the platform (Fig. 4.38).

The connection of the gable slender structure to the pipes is realized with collars and the full structure itself allows an easy positioning of cables and inverters. The wide space under the panels can be utilized to shelter electric devices without the necessity of sacrificing the external space for auxiliary systems. One of the important findings of the Singapore test bed was the fact that panels perform best if the ventilation is good and that free space allows the air exchange needed. Furthermore an important aspect of the gable structure is its compactness. The gable raft occupies a surface of about 60 m^2 and with PV modules of 400 W a FPV plant of 1.6 MWp occupies only 1 ha of the basin surface.

Finally the cost of gable slender should be less than 20 c\$ per Watt so this solution should be very competitive with other floating solutions and even with land-based PV structures.

Gable slender is naturally fit for low latitudes but recent studies demonstrate that even at high latitudes where diffused light is predominant on direct light the overall harvesting of such a design is not far from what we can get from a standard system.

Actually, results for yearly energy yield are shown in Fig. 4.39: blue dots represent the gain due to the cooling effect and red dots, the gain/loss with respect to a fixed plant south oriented with optimal tilt.

Gable slender is naturally fit for low latitudes but recent studies demonstrate that even at high latitudes where diffused light is predominant on direct light the overall harvesting of such a system is not far from what we can get from a standard system.

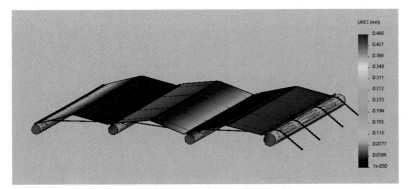

FIG. 4.36 Two rafts connected with hinges.

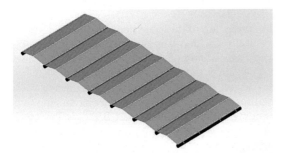

FIG. 4.37 Seven rafts connected on the long side.

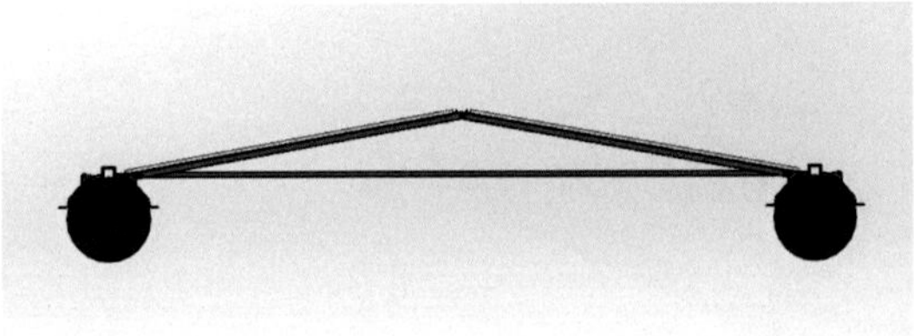

FIG. 4.38 Vertical section of gable slender.

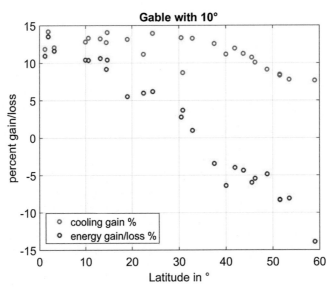

FIG. 4.39 Gain/Loss using Gable slender for 25 towns between Singapore and Stockholm.

REFERENCES

[1] Y. Lee, H. Joo, S. Yoon, Design and installation of floating type photovoltaic energy generation system using FRP members, Solar Energy 108 (2014) 13−27.

[2] Y. Choi, A study on power generation analysis of floating PV system considering environmental impact, International Journal of Software Engineering and Its Applications 8 (2014) 75−84.

[3] CNS_LG, 2018. [Online]. Available: https://www.lgcns.com/Views/News/NewsDetail?SERIAL_NO=1560.

[4] T. Jongsma, S. Eggink, S. Kreiter, Floating Bifacial : reflection on power, in: BifiPV Workshop, Konstanz, 2017.

[5] 4C-Solar [Online]. Available: https://www.4csolar.com/lakes.

[6] T. Reindl, At the Hearth of Floating Solar: Singapore, February 2018, pp. 18−23. www.pv-tech.org.

[7] M. Sumimoto [Online]. Available: https://tech.nikkeibp.co.jp/dm/atclen/news_en/15mk/051401330/?SS=imgview_msbe&FD=48575398.

[8] HDB [Online]. Available: https://www.straitstimes.com/singapore/environment/hdb-exploring-floating-solar-panels-in-open-sea.

[9] [Online]. Available: https://adtechindia.com/solar-energy/floating-solar/.

[10] NRG-Energia [Online]. Available: http://www.nrg-energia.it/floating-pv-systems.html.

[11] Takiron [Online]. Available: https://tech.nikkeibp.co.jp/dm/atclen/news_en/15mk/032201221/?ST=msbe&P=2.

[12] Swimsol [Online]. Available: https://swimsol.com/solar-projects/floating-photovoltaic-offshore-solar-sea-power-pv-modular-96kwp/.

[13] Solaris-Floating [Online]. Available: https://www.solarisfloat.com/.

[14] Nemo-Eng [Online]. Available: http://nemoeng.com/eng/surface/.

[15] Moss-Marine [Online]. Available: http://www.mossww.com/.

[16] R. Cazzaniga, M. Cicu, M.-C. Rosa-Clot, C. Ventura, Floating photovoltaic plants: performance analysis and design solutions, Renewable and Sustainable Energy Review (2017).

[17] M. Rosa-Clot, G.M. Tina, Submerged and Floating Photovoltaic Systems, Modelling, Design, Case Studies, Elsevier, Academic Press, London, 2017.

[18] R. Cazzaniga, M. Rosa-Clot, P. Rosa-Clot, G. Tina, Floating tracking cooling concentrating (FTCC) systems, in: 38th IEEE Photovoltaic Specialists Conference (PVSC), Austin (USA), 2012.

CHAPTER 5

Environmental Loads, Motions, and Mooring Systems

TREVOR WHITTAKER, PHD, BSC, FRENG., FRINA, FICE, CENG •
MATT FOLLEY, PHD, BSC • JONATHAN HANCOCK, MENG (HONS), PHD, MICE, CENG

1. INTRODUCTION

By definition floating solar power plants will have a very different support structure to their land-based versions. However, this difference is not only in the requirement that their supporting structure floats but also because the environmental forces that they experience will be different and that factors such as dynamic response and station-keeping requirements must also be considered. The latter is not generally part of the design of land-based systems.

The environmental loads that are common to both floating and land-based solar power plants are wind loads, self-weight, maintenance/access loads, and in some locations snow loads. However, in land-based systems it may only be the maximum wind load that is considered in the design, whereas in a floating system it is also important to consider both the spatial and temporal distribution of the wind loads. The requirement to consider the spatial distribution of the wind loads arises for the recognition that there is likely to be far less connections to the ground through moorings in a floating solar power plant due to their cost. In addition, the net load on these anchors will depend on the coherence of the wind loads across all of the panels restrained by a particular anchor. The temporal distribution of the wind loads is important because these will excite the dynamic response of the system, and the motion of the panels needs to be known to ensure that the structure can cope with these induced motions.

In addition to wind loads, floating solar plants will also be subject to loads due to waves and also due to water currents. Waves can exist on any body of water and are generated by the wind blowing across the water surface, which starts by generating ripples that grow

over time and distance into larger waves. Currents are associated with the net transport of water and are most obvious in rivers as the water flows downstream. However, they may also be significant in reservoirs close to intake and outtake points as well as around coasts due to tides or other marine currents. As with wind loads, the temporal and spatial characteristics of the wave and current loads need to be considered in the design of floating solar power plants.

A comprehensive treatment of any one of these environmental loads would fill a book in its own right. Accordingly, this chapter can only provide an initial overview of these loads to allow the reader to understand their fundamental characteristics and their potential impact on the design of a floating solar power plant. To support this, this chapter also provides details on how the dynamic response of a floating solar power plant may be modeled, together with the specification of key aspects of the dynamics that may influence the system design. In addition, this chapter considers the mooring systems for different types of floating solar power plant as they have a significant influence on the dynamic response of the array subject to all the environmental loads encountered. This chapter finishes with a bibliography for readers who are interested in studying these aspects of the design of floating solar power plants in more depth.

2. WAVES
Physics of Waves

A very good general description of gravity water waves can be found in Chapter 1 of the Open University course book on *Waves, Tides and Shallow Water Processes* [1]. As floating solar plants are deployed in larger

Floating PV Plants. https://doi.org/10.1016/B978-0-12-817061-8.00005-1

47

bodies of water such as large lakes, inland seas, and coastal regions, waves become a much more significant aspect of the design for survival in storm conditions. In order to appreciate the challenge of designing for more energetic sea states, it is important to understand the basic physics of wind-generated water waves. A water wave is not all that it seems to the observer seeing the undulation of the water surface and the appearance of forward motion of the wave crests. Wave action is due to energy imparted to the water by wind blowing over its surface and although the wave has an obvious direction of travel the fluid itself does not move bodily from A to B but in fact only moves locally. As shown in Fig. 5.1, in deep water, the fluid particles beneath the surface move in circular orbits, the radius of which decay exponentially with depth such that at a depth equivalent to half the wavelength the radius is only 3% of that at the surface. As the water depth reduces to less than half a wavelength, the fluid particle orbits become more elliptical with the vertical component of the motion reducing, as shown in the diagram for intermediate and shallow water. Consequently, when a floating body is placed in a wave it reacts not only to the surface motion but also to the motion of the fluid particles beneath. Therefore, the body is acted on by both the potential and kinetic energy in the waves. Small amplitude sinusoidal waves, as described above are the basic building blocks from which the more complicated descriptions of "real seas" are built. Observation of wind-blown waves in open water does not show nice simple single frequency sinusoidal waves but a much more complex conglomeration of wave periods and directions of travel.

Wave Generation

Anyone that has looked at the surface of a body of water such as a pond or swimming pool on a windy day will have noticed that small waves or ripples are generated that travel in the direction the wind is blowing. It is perhaps not so obvious that this same action of the wind blowing across the surface of the ocean is responsible for the giant surfing waves of Hawaii. The only fundamental difference between these two waves is the duration in which the waves are growing due to the wind. If it is assumed that the wind is blowing constantly, then the duration that the waves can grow depends on the length of water in the direction from which the wind is coming from. This length is sufficiently important in the generation of waves that it is given its own term—*Fetch*. The other fundamental factor that is important in the generation of waves is the wind speed, which when combined with the fetch can be used to define the characteristics of waves. A typical Nomogram which might be used for deep water offshore sites with fully developed seas is given in Fig. 5.2 from the Shore Protection Manual (1984), SPM. By way of illustration consider the following example a site is has a fetch 2 km long in the North-South direction and 1 km wide in the East-West direction. Consider a constant wind blowing from North to South at a speed of 20 m/s. In this case the largest waves will occur on the South shore with a significant wave height of approximately 0.65 m. However, the same wind blowing from East to West would result in the largest waves being on the Western shore with a significant wave height of approximately 0.45 m. The nomogram also shows the substantial increase in significant wave height with larger fetch distances and greater wind speeds. However, such a nomogram should only be used as a rough first approximation of wave climate. A full numerical analysis should be undertaken for each site under consideration. This is because there are other factors which limit wave height such as shallow water and white capping which occurs at higher wind speeds and is recognized by the tops of the wave crests being blown over. Other Nomograms exist which are better

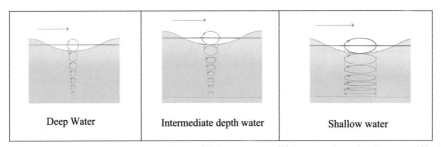

FIG. 5.1 Wave-induced water particle motions. **(A)** Deep water; **(B)** intermediate depth water; **(C)** shallow water.

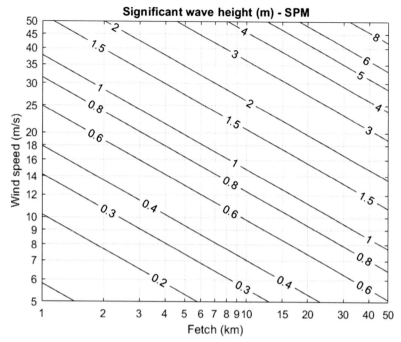

FIG. 5.2 Relationship between wave height with fetch and wind speed from the *Shore Protection Manual* [2].

suited to the restricted bodies of water where most floating solar plant have been deployed to date. For example an alternative nomogram derived from the JONSWAP spectrum predicts wave heights around 30% lower than those in Fig. 5.2.

It may be noted that the term significant wave height is used to define the waves. This term is used because the waves generated by the wind will not have the same height but are continually varying. The significant wave height was originally defined as the average height of the waves as estimated by an experienced observer. Then, as measurements became more common, it was first redefined as the average height of the third highest waves and more recently redefined again as four times the standard deviation of the surface elevation. Although these different definitions can in some circumstances result in slightly different estimates of the significant wave height, it is essentially a measure of the size of the wave.

Wave Spectrum

However, greater wind speeds and fetch distances also result in longer wave lengths and increased peak spectral wave periods. To understand the definition of the peak spectral wave period, it is necessary to first consider how the variability of the waves in "real" sea states is described. It is common to represent this variability using a frequency spectrum of wave energy associated with the surface elevation at each frequency so that the energy in each frequency band is proportional to the area in each segment, as shown in Fig. 5.3. Wave frequency is the reciprocal of wave period so that longer period waves have a lower frequency. A common understanding of a frequency spectrum is that the surface is generated by the summation of multiple sinusoidal wave components, each with a different frequency and with an amplitude defined by the energy at each wave's frequency and with a random phase relationship to all the other waves in the spectrum.

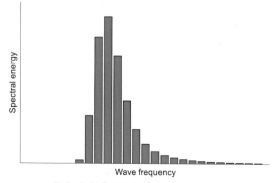

FIG. 5.3 Example of wave spectrum.

Although this is not an entirely accurate interpretation of wind-generated waves, it is surprisingly good and can be used in virtually all conditions except for breaking waves, where nonlinear coupling between the wave components complicates the representation. The peak spectral wave period is the wave period associated with the maximum variance density. In the spectra, the left-hand side of the curve is determined by the maximum energy transfer from wind to wave (wave generation) while the right-hand side is governed by the maximum wave steepness before the wave breaks (breaking wave limit).

A number of different spectra are available that define the relationship between the spectral variance density and the wave frequency. A commonly used spectra, which is suitable for a developing sea state, is the JONSWAP spectrum. Fig. 5.4 shows the JONSWAP spectrum for a 20-m/s wind with different fetch lengths (or wind durations), which shows that as the sea state grows the spectral peak shifts to lower frequency and the total spectral variance increases. The series of curves from right to left shows how a sea state develops from low energy high frequency or short period waves to more energetic shorter frequency longer period waves. Therefore, small ripples with increasing time and fetch distance ultimately become large energetic ocean swells as seen in the North Atlantic as weather systems travel from the east coast of North America to the western shorelines of Europe.

In reality, wind rarely blows at a constant velocity, and variations in the wind speed and direction result in more complex spectra. An important characteristic

of waves is that they persist after the wind stops blowing, continuing the travel in the direction from which they have been generated. Thus, in large bodies of water, waves can leave the area where they are generated and travel to other regions that may be a long distance away from their source. For example, waves that have been generated in the South Atlantic can propagate to the North Atlantic, where they impact the coastline of North Africa. These waves can combine with waves that may be generated locally or from somewhere in the North Atlantic to produce a complex wave spectrum that does not only vary with frequency but also with direction. An example of a directional wave spectrum is shown in Fig. 5.5, where waves from at least two different sources can be clearly identified (a wind sea and swell). In this plot, the wave color indicates the spectral energy density, while the frequency of this energy is defined by the distance from the center of the circle.

Influence of Water Currents and Water Depth on the Wave Spectrum
The directional wave spectrum is also influenced by water currents and water depth, both of which have the potential to change the wave frequency and the wave direction. These processes are complex and typically require a numerical model to provide an accurate estimate of the change in wave spectrum. Notwithstanding, it is possible to make some general estimates of their potential by considering the underlying factors that influence their effect on the waves.

Water currents change the observed wave frequency due to doppler shifts and also change the wave direction

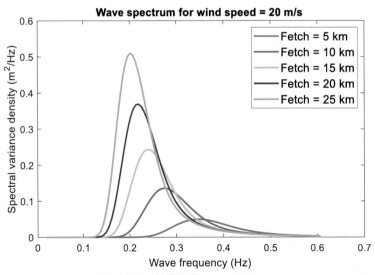

FIG. 5.4 JONSWAP wave spectrum.

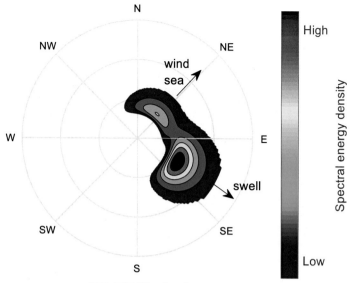

FIG. 5.5 Directional wave spectrum.

as the waves will tend to align with the direction of the water currents. Conservation of energy also means that these currents can cause a change in the wave height as the wave energy can be concentrated or dispersed by the currents. Thus, where there is an adverse current to the direction of wave propagation the speed of wave energy, propagation will reduce and so the density of wave energy will increase leading to an increase in the wave heights and in extreme cases to wave breaking. This is evidenced by wave conditions known as overfalls off headlands and at the entrance to channels where ocean swells meet opposing currents.

The water depth affects the speed of wave propagation through a process called shoaling when the water depth is less than about half a wavelength. The speed of wave propagation reduces in shallow water, and conservation of energy means that this causes the wave heights to increase, at least until the waves break. However, where the direction of wave propagation is not orthogonal to the depth contours, the bathymetry can also change the direction and amplitude of the waves through a process called refraction. Wave refraction can be explained by considering a long wave crest approaching a coastline. If the crest and the bed contours are not aligned initially, one end of the crest will slow first due to shoaling thus slowing down while the other end is still moving faster. Thus, the crest changes its direction of travel so that the wave crests will tend to align with the bottom contours. This is the reason why waves are typically approximately parallel to the shoreline when they reach the beach. Fig. 5.6 shows that wave refraction concentrates the wave rays off headlands increasing the wave energy and height while reducing it in bays.

A potentially important impact of water currents and water depth is the generation of "hot" and "cold" spots where the waves are larger or smaller than the surrounding conditions. This could clearly have important implications for the siting of a floating PV plant, as a location with smaller waves is likely to experience smaller wave loads. However, care must be exercised in the exploitation of these localized conditions as they are generally dependent on the incident directional wave spectra, which are rarely consistent. Also potential changes in the bathymetry due to movements of sands/mud can change the water depth and

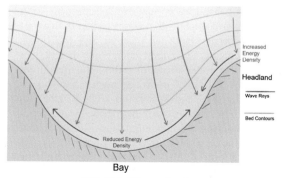

FIG. 5.6 Wave refraction.

thus the temporal location of "hot" and "cold" spots should also be considered.

Extreme Waves

Identifying the largest wave that is likely to occur at the deployment location of the floating photovoltaic (FPV) plant is an important element of the design specification. In a relatively small body of water, such as a reservoir or inland sea, which is only affected by a single wind field, then the largest theoretical significant wave height and associated peak wave period can be estimated using an appropriate approximate method such as a Nomogram based on the maximum wind speed and fetch length. Large deployments in larger bodies of water may require a more detailed analysis that considers the spatial and temporal variation of the wind field across the body of water together with hindcast models that allow the largest significant wave height and peak period at a particular location to be estimated based on historical data. Typically, a minimum of 10 years of data are assumed to be required to provide a relatively stable estimate of the wave climate. Extreme value analysis can then be used to estimate the extreme sea state that can be expected to occur once every 50 or 100 years (the 50-year or 100-year return sea state). However, it is important to recognize that whether the Nomogram or the wave climate is used, what is predicted is not the largest wave, but the largest significant wave height and associated peak period.

The standard definition for the significant wave height is that it is "the average height of the third highest waves." From this definition, it is possible to make an estimate of the largest wave height, which is expected to be approximately twice the significant wave height. For example, the largest wave height in a sea state with a significant wave height of 2.0 m is expected to be approximately 4.0 m. It is clearly important that there is no confusion between the maximum and significant wave heights as this could lead to serious underdesigning of the structure.

A final consideration for the specification of a maximum wave height is that waves greater than about 80% of the water depth are unstable and will break, thereby losing some of their energy and height. For example, in a water depth of 5.0 m, the maximum wave height will be 4.0 m. Waves bigger than this, as predicted by the wind speed and fetch length, will have broken before they reach the specified water depth and thus limiting the maximum wave height at that location. The maximum steepness of waves is also limited to about 1/7, that is the wave height must be less than about 1/7th of the wavelength, which provides

an additional constraint on the wave height in deeper water. However, it is possible to exceed the breaking wave height limit if wave crests from different locations combine at a specific location.

Wave Loads

The wave loads on a floating structure can be characterized as being the combination of one or more of the forces due to diffraction, drag, and inertia. Diffraction (and radiation) forces are important when the structural volume is sufficiently large to create waves, which becomes important when the dimensions are larger than about half a wavelength (λ). If the wave heights (H) and body motions are small relative to the body dimensions (D), then the effect of drag forces can become negligible so that inertial forces are dominant. Fig. 5.7 illustrates the different wave force regimes that are important to consider based on the wave height and wavelength relative to the body dimensions.

Diffraction forces arise from the reflection and radiation of waves caused by the presence and motions the body. Although analytical solutions exist for the diffraction forces for idealized bodies such as spheres, these forces are generally estimated using a linear potential flow solver such as WAMIT (www.wamit.com/). Linear

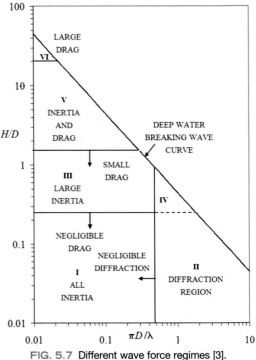

FIG. 5.7 Different wave force regimes [3].

potential flow solvers calculate the frequency-dependent wave force coefficients, which define the forces per unit wave amplitude or body motion. Because these forces are linear (proportional to the wave amplitude and body motion) it is possible to calculate the total diffraction force as the sum of forces due to the individual wave components in the wave spectrum.

The drag and inertia forces are defined as the two contributing components in the semiempirical Morrisons equation (see for example [4]). By definition, the drag force is in phase with the instantaneous relative flow velocity U and proportional to the relative flow velocity squared, with the drag force always acting to oppose the relative motion. Also, by definition, the inertial force is in phase with and proportional to the instantaneous relative flow acceleration $\dot{U}$. Thus, the Morrisons equation is given by

$$F = \frac{1}{2}\rho C_D A U |U| + \rho C_A V \dot{U}$$

where ρ is the water density, A is a representative cross-sectional area, V is the body volume, and C_D, C_A are the drag and inertia coefficients, respectively. Values for the drag and inertia coefficients in the Morrisons Equation are typically obtained from experiments. In the majority of cases, experimental data for the exact shape will not exist. However, typically an experienced designer is able to make a reasonably good estimate of the drag and inertia coefficients by making some simplifying assumptions. In the cases where relevant experimental data do not exist, then it is usually required that appropriate wave tank testing is undertaken to produce estimates of these coefficients, with particular consideration given to potential scaling issues.

Wave Slamming

An additional source of infrequent wave loads on a floating structure can be due to wave slamming. A wave slam will occur when there is a sudden retardation of a volume of water, such as when a plunging wave breaks onto a flat surface, or when the water surface hits the underside of a flat surface. The rapid deceleration of the water can result in a considerable force on the structure. Consequently, it is common to design the structure so that any flat surfaces are sufficiently high above the water surface to eliminate the possibility of the water reaching the flat surface, or to avoid the use of large flat surfaces so that there is never a rapid deceleration of the water. This is obviously a challenge for floating solar PV arrays which tend to have low freeboard and as they are deployed in more exposed wave

locations. In a large array, the panels which are facing the approaching waves are the most vulnerable to wave slam, and it may be necessary to protect the leading panels by setting them back from the edge and placing structure in front to take potential impacts. Even panels within the body of the array could be susceptible to wave slam from below as the structure will tend to follow the vertical motion of the water surface with potentially relatively little attenuation of the wave as it passes beneath.

3. WIND
Characterization of Wind

Wind is generated by the differential heating of the earth's surface, which leads to a pressure gradient and then air moving from areas of high pressure to areas of low pressure. Although it is common to refer to a particular constant wind speed and direction, in actuality the wind is continually changing with time due to gusts and turbulence. Consequently, wind speed data are typically characterized by a temporal mean magnitude and direction. The standard deviation of the wind speed during this time is a measure of the variability of the wind speed and defined as the turbulence or gustiness. The ratio of the standard deviation of the wind speed to the mean wind speed is termed the turbulence intensity. The most common duration used for defining the mean wind speed and direction is 10 minutes because this provides sufficient data to obtain a relatively stable estimate of the mean and in most circumstances the wind can be considered to be statistically stationary for this duration.

The wind speed is also typically defined at a particular height above the ground or sea surface. A common reference height to use for the wind speed is 10 m because in most circumstances the wind at this height is not strongly affected by local sources of disturbance such as trees and buildings, which means model predictions of the wind speed at 10 m above the ground are relatively accurate. However, what is typically required is the wind speed close to the height of the solar panels, which is likely to be much less than at 10 m elevation. In this case, the wind speed at the height of the solar panels can be calculated by assuming a particular wind profile, although clearly measuring the wind speed at the appropriate height is the preferred approach if this option is available.

A number of different empirical wind profiles, which provide relationships between winds speeds at different heights, have been proposed with the most common being the logarithmic profile model, the power law model, and the Frøya model. All of these models have additional parameters that modify the wind profile based on local conditions, including the proximity of buildings and trees on land and including the presence of waves above water, which define a roughness parameter. Essentially, the smoother the surface over which the wind blows, the smaller the roughness parameter. Examples of possible wind profiles for three different *roughness parameter* values are shown in Fig. 5.8 for a 10 m/s wind at 10 m above the ground. It can be seen that depending on the *roughness parameter* the wind speed at 1 m above the ground could vary from about 6.5 to 8.0 m/s. Thus, converting a wind speed from the reference height to a wind speed at the height relevant for the design of the structure requires determining a suitable *roughness parameter* and calculating the expected wind profile.

While the expected mean wind speed, direction, and standard deviation may be sufficient for the design of a land-based PV plant (as these can be used to estimate the maximum wind speed), the temporal and spatial variations in the wind field are also likely to be important for floating PV plants due to the significance of the wind spectrum and spatial coherence of the wind speed on the design of floating structures.

The wind spectrum is important because it defines the frequency content of the wind loading that will influence the dynamics of the floating structure, which could result in significant structural damage if not adequately accounted for in the design. The wind spectrum, similar to the wave spectrum (see Wave Spectrum section), is a representation of how the wind speed varies in time. The magnitude of the spectrum at a particular frequency indicates how much the wind speed varies at that frequency. Ideally data should be collected that enable the frequency content of the wind to be reproduced accurately. However, in the absence of measured data, several empirical model spectra exist to represent the frequency content of the wind speed. An example of a wind spectrum, the Harris spectrum [5], is shown in Fig. 5.9. This shows that the largest variations in wind speed occur at the lowest frequencies and that as the frequency of variation increases the amount of variation in the wind speed reduces. This can be understood by recognizing that the inertia in the air means that large high frequency variations in the wind speed are not possible.

All the empirical model spectra are scaled by an *integral length scale*, which is a measure of the size of the turbulent structures in the wind. Small obstructions in the wind will typically cause small turbulent structures and so result in a small *integral length scale*. Conversely, large obstructions will typically cause large turbulent structures and so result in a large *integral length scale*. Various expressions exist for estimating the *integral length scale* that are functions of the *roughness parameter* and the height above the ground or sea surface.

The spatial coherence is important because it defines the expected difference of the instantaneous wind speed

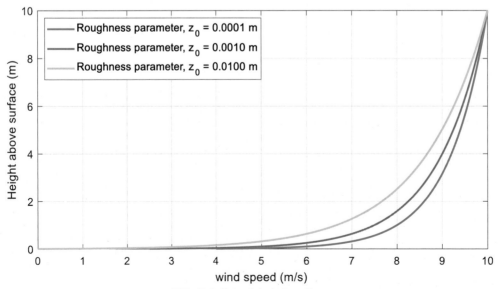

FIG. 5.8 Potential wind profiles.

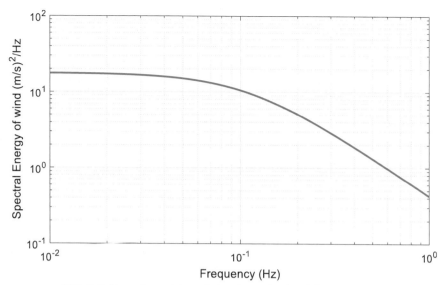

FIG. 5.9 Example model spectral variance density of the wind speed.

at two different locations, which will influence the net load on the floating structure. The spatial coherence of the wind speed from one point in space to another can be defined by the correlation between their respective wind speeds. This correlation can be expressed as a coherence spectrum, which is a frequency and distance-dependent measure of the spatial correlation of the wind speed. A high correlation between two points means that the wind speed tends to increase and decrease at both points at the same time (a correlation of 1.0 would mean that the wind speed variations are completely correlated and so increase and decrease in tandem). Conversely, a low correlation between two points means that the wind speed variation at one point occurs largely independently of the wind speed variation at the other point (a correlation of 0.0 would mean that the wind speed variations are totally independent of each other). As for the frequency spectrum, it is preferable that the spatial coherence spectrum of the wind field is determined through measurement, but a model coherence spectrum can be used in the absence of these data. Again, a number of empirical model coherence spectra exist to represent the spatial coherence of the wind field. The exponential IEC coherence spectrum for a wind speed of 25 m/s and an *integral length scale* of 100 m is shown in Fig. 5.10 for the along-wind correlation. For example, for the conditions shown in Fig. 5.10, the correlation between 0.1 Hz variations in the wind speed for two points separated by 3 m is approximately 0.8, which is relatively high correlation, and so the wind at these two points will tend to

vary in unison. However, the correlation between 0.2 Hz variations in the wind speed for two points separated by 10 m is approximately 0.25, which is a relatively low correlation and so the wind at these two points will tend to vary independently.

As would be expected, the coherence between two points decreases as both the separation and the frequency increase. In addition (not shown), the coherence tends to decrease with the wind speed, so that at lower wind speeds there is less coherence between the wind speeds at two locations.

Extreme Winds

The expected extreme mean wind speed for a particular site can be estimated using the long-term distribution of the wind speed. In hurricane-free areas the two-parameter Weibull distribution can be used for this distribution, where the scale and shape parameters of the Weibull distribution depend on the site conditions. Specifically, the scale parameter is a measure of the average mean wind speed, and the shape parameter is a measure of the mean wind speed variability. An example of a Weibull distribution is shown in Fig. 5.11 for an annual average wind speed of 6.2 m/s. The Weibull distribution can then be used directly to estimate the expected 50-year 10-minute mean wind speed, which is equivalent to a 0.4×10^{-6} probability of occurrence and so equates to a 10-minute wind speed of approximately 32 m/s. Where hurricanes occur, the wind speed distribution for the estimation of extreme mean wind speeds should be obtained

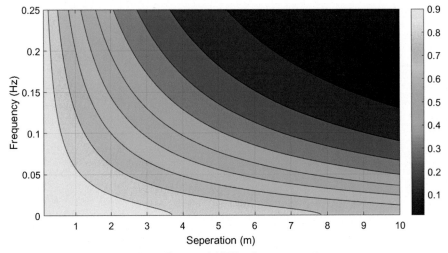

FIG. 5.10 Exponential IEC coherence spectrum.

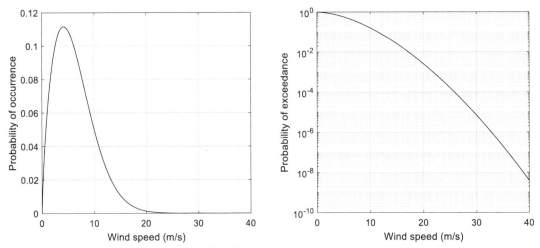

FIG. 5.11 Weibull distribution of mean wind speed.

using the available hurricane data because the Weibull distribution is typically not suitable for these circumstances.

For calculating wind loads, it is often necessary to estimate the maximum wind speed for durations that are shorter than 10 minutes. This can be achieved using the Durst [6] Curve or similar, which provides an empirical relationship between the maximum wind speeds for different durations. An example of the Durst Curve is provided in Fig. 5.12. The ratio of maximum speeds for two different durations can be calculated by taking the ratio of the two factors. For example, if the maximum 10-minute (600 seconds − factor = 1.05) average speed for a site is 25 m/s, then the maximum 3-second (factor = 1.5) average speed will be 25 × 1.5/1.05 ≈ 36 m/s.

Wind Loads

Structural loads due to the wind can be due to the dynamic pressure, associated with the reduction in the normal wind speed to zero orthogonal to the surface, lift forces due to a variation in wind speed over the surface of the body, shear forces due to viscous boundary layer forces, and vortex shedding due to flow separation from the body. In general, all bodies are likely to be affected by all of these forces, but typically one of these forces will be dominant and likely to dictate the structural wind loads.

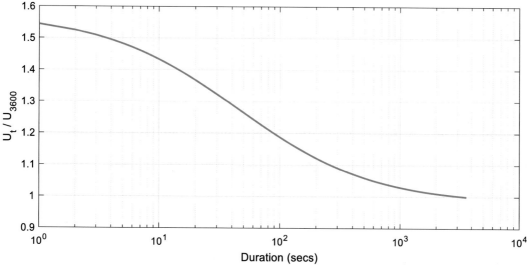

FIG. 5.12 The Durst Curve.

Dynamic pressure loads will typically be dominant on large flat surfaces. If you have ever tried to carry a large flat board on a windy day you will be familiar with this force. The dynamic pressure of a fluid is proportional to the velocity squared and so doubling the wind speed is likely to result in a quadrupling of the dynamic pressure load. In addition, dynamic pressure loads tend to increase faster than the area of the flat surface because the pressure generated is related to the obstruction of the wind. Thus, providing small vents in a large flat surface can result in a dramatic decrease in the dynamic pressure load as air can escape through these vents and thereby reducing the pressure. In addition, a decrease in the spatial coherence of the wind speed is also likely to reduce the maximum dynamic pressure load as the pressure on the surface will not all act at the same time. However, it is important to note that a large surface is likely to have an influence on the wind speeds and their spatial coherence and so detailed modeling or testing is likely to be required to produce a reliable estimate of dynamic pressure loads for large surfaces.

Lift is the force that supports aeroplanes in the sky and so can be very large. In this case, aerofoils are especially designed to generate large lift forces, but they can be generated by any body where the surface wind speed differs for opposite sides of the body. However, the extent to which lift forces are generated on a body depends on the characteristics of the body boundary layer and whether the flow separates from the body. The effect of flow separation is known as stall, where the lift force reduces rapidly with the onset of flow separation. Thus, the accurate estimation of lift forces is extremely challenging as it depends both on the large-scale flows around the body as well as the small-scale flows close to the body surface that dictate whether the flow separates from the body. In many cases reference to physical modeling results are required to make a reasonable estimate of the lift forces.

Viscous boundary layer forces and vortex shedding forces are both associated with the wind flowing past a body and although they are fundamentally different, with different characteristics, they are often combined to produce a single load source. This is partially a pragmatic solution because it reduces the complexity of the calculation, but also because it is not simple to separate the two forces as the vortex shedding will influence the shear velocity and thus there is coupling between these two forces that cannot be entirely separated. This load is also generally proportional to the square of the wind speed and influenced by the shape of the body and its proximity to other bodies that are either adjacent to it, or in-front/behind it.

Slender bodies can also be influenced by the shedding of vortices on alternate sides to produce what is known as a Kármán Street Vortex. The creation of a Kármán Street Vortex depends on the Reynolds number, for which a limiting value of about 90 (based on the cylinder diameter) is required before the street starts to form. The shedding of these forces results in an alternating force that is orthogonal to the direction of flow and the axis of the slender body. Support struts and mooring lines are common structural elements that can be affected by Kármán Street Vortex—induced loads. The frequency of the alternating force also depends on the Reynolds number and can be important when considering the dynamics of the structure (see Analysis of System Response Section).

4. WATER CURRENTS
Source of Water Currents

The source of water currents that may affect an FPV plant will depend on where it is located. With respect to the types of water currents, potential locations for an FPV plant include on a river, in a reservoir, and in the sea, each of which is affected by a different source of water current. In addition, there are locations that may be affected by more than one source of water current, such as estuaries and lakes with significant inflows and outflows of water. In these cases, the combined effects of the different sources of water currents need to be considered.

The water currents in a river are primarily due to the flow of water downstream. However, meanders in a river can cause eddies and recirculation patterns that can result in local flows that go against the bulk flow of the water. Generally, the more complex the course of the river, with sharper and more extensive bends, the more complex the flow patterns. For a given water flow rate these patterns are likely to be relatively stable and predictable. These patterns are also likely to be dependent on the bulk flow rate of the river and so may vary throughout the year as changes in rainfall cause changes in the amount of water flowing downstream.

The water currents in a reservoir will be primarily due to the filling and emptying of the reservoir. Away from intakes and outtakes the water currents are likely to be small, but close to these points relatively high-speed water currents may occur. Obviously, the magnitude of the water currents will depend on the rate of filling and emptying of the reservoir, which will vary with the time of year as well as potentially other roles of the reservoir, such as irrigation, power production, and/or flood control.

Away from the mouths of rivers, the primary water currents in the sea are likely to be due to the tides or wind-driven surface currents. The water currents due to tides are typically very regular with an approximately 12.5-h cycle that increases and decreases in magnitude twice every lunar cycle (approximately every 29.5 days). Although the tides are relatively predictable, along complex coastlines or in estuaries, the tide-induced water currents can be complex. In addition to

tides, a number of marine currents associated with ocean circulation patterns also exist. The characteristics of these ocean circulations patterns are generally related to larger global meteorological conditions and their effect specific to a particular site.

As noted, some locations will be affected by more than one source of water current. For example, the water currents in an estuary will be affected by the river flow downstream and the tides. Unfortunately, the resulting water current is unlikely to be the simple addition of these two currents, but they are likely to interact in a complex fashion to produce a potentially complex flow pattern.

Current-Induced Loads

The main sources of structural loads due to water currents are similar to those for wind loads; dynamic pressure, associated with the reduction in the normal current speed to zero orthogonal to the surface, shear forces due to viscous boundary layer forces, and vortex shedding due to flow separation from the body, together with an additional structural load due to the wave-making resistance of the body. However, because the water velocities are much smaller than those of the wind, the dynamic pressure effects tends to be less important (as these are proportional to the velocity squared) and the viscous boundary layer forces (proportional to velocity) tend to be more important. The load due to the wave-making resistance of the body is related to the energy in the waves generated when diverting the water around the body and so depends on the shape of the body as well as its displacement.

Where there are also significant wave-induced velocities then the current velocity should be combined with the wave-induced current velocities to produce an estimate of the total local velocity. This is particularly important for small bodies where the current-induced load is typically proportional to the square of the relative velocity because it is generated by vortex shedding.

5. SYSTEM DYNAMICS
Numerical Modeling

In cases where a body can be considered as stationary, then the environmental loads can be used directly to calculate the stresses in the structure. However, an FPV plant cannot be considered to be a stationary structure and so it is necessary to define the equations of motion for the structure, which can subsequently be used in a numerical model to determine the system dynamics. The equations of motion can be constructed using Newton's Second Law, which states that a body's acceleration is proportional to the net force acting on the body and inversely proportional to the body's mass. An FPV plant is likely to consist of multiple bodies, and Newton's Second Law can be applied to each of these bodies to define its equations of motion. To solve Newton's Second Law for each body, it is necessary to know the environmental loads that act on the body, as well as loads through connections to other bodies and any mooring forces that act directly on the body.

In general, the number of equations of motion or degrees of freedom required to completely define the response of the system is equal to six plus the number of independent joints in the structure plus any equations of motion associated with the moorings. The 6° of freedom relate to the whole system movement in the three translational and three rotational degrees of freedom. For example, a plant that consists of 20 rigid floats, each connected to at least one other float using a hinge joint, would have a total of 25 degrees of freedom (ignoring any associated with the moorings). The state of each degree of freedom can be defined by its displacement and its velocity (or rotational equivalents), so that the number of variables required to define the system is 12 plus twice the number of independent joints. So, in the example of the plant that consists of 20 hinged floats, the equations of motion that is required to fully define its motion would have 50 independent variables.

Flexible structures present a potentially more complex case as the structural deformation could theoretically adopt a multitude of different shapes. The equation of motion for a flexible structure is typically defined either by discretizing the structure into n rigid bodies or by defining m mode shapes that can be superimposed to produce an approximation of the potential shapes of the flexible structure. Essentially the larger the value of n or m, the more accurate the model of the flexible structure becomes. However, this must be balanced against the increase in the computational effort in solving the equations of motion which also increases with the value of n or m.

It is important to include the forces on the bodies due to moorings in the system model. This could be as simple as including a horizontal spring term that stops FPV plant from drifting away from its station plus an additional set of equations of motion that provide an accurate representation of the mooring system. Typically, the requirement for a more accurate representation of the mooring system increases with the inherent nonlinearity of the moorings and the potential for complex mooring dynamics. This needs to be balanced by the computational resources available and the development phase of the system model. That is, a simple model of the moorings may be sufficient to support design during the early phases of development, while a more complex model may be required during the final design phase.

Once all the equations of motion for the system have been defined, it is possible to generate a suitable numerical model. There are two basic methods for solving the equations of motion; time-domain models or frequency-domain models. A time-domain model calculates the response of the system at consecutive discrete time steps, with the results of the previous time step used to inform the conditions for the next time-step. A frequency-domain model assumes that the response of the system can be represented as the summation of a set of sinusoidal oscillations in each of the degrees of freedom. Typically, both types of model would be expected to be used in the design of an FPV plant. A time-domain model can provide accurate estimates of the expected motions and loads in the system, which can be used to assess whether the design is fit-for-purpose. However, a time-domain model is typically computationally expensive, especially for a system with multiple degrees of freedom. On the other hand, a frequency-domain model is computationally less demanding and is capable of identifying potentially problematic issues in the design that can be defined by response amplitude operators (RAOs) and resonant frequencies (see Analysis of System Response section).

As well as whole system numerical models, it can also be useful to produce models of particular subsystems that would benefit from more detailed analysis. Specifically, these subsystem models could provide valuable information directly or parameterized to provide appropriate characterization to be used in the whole system numerical model. An example of a subsystem that could be modeled is a single float. The results of this analysis could help to identify any particular load concentration or hydro-elastic interactions or provide better estimates of added mass and drag coefficients to be used in the whole system model. Importantly, a highly computationally demanding technique such as Computational Fluid Dynamics (CFD) could be used for this, which would not be practicable for modeling the whole system due to limitations in the available computational resource.

Physical Modeling

The production of a numerical model of the system requires making a number of assumptions and approximations with a range of accuracies. Where sufficient knowledge and experience has been developed in the application of a particular numerical model it may be possible to consider the results reliable without additional validation. However, in many cases the design of FPV plants is likely to include elements for which there is limited knowledge and experience, which increases the uncertainty in the model and the risk in the application of the modeling results. Physical modeling can be used to reduce the uncertainty and risk by providing calibration or validation of the numerical models, as well providing relevant design data (such as maximum expected loads) directly.

Physical modeling typically involves constructing a scale model of the proposed plant and testing this in a wave tank, where the waves are generated by a set of computer-controlled paddles. An example of this type of facility is the QUB wave tank at Portaferry, which is shown in Fig. 5.13. Air blowers may also be used to simulate the effect of wind and currents can be created in some wave-tanks to simulate the effect of water currents. Alternatively, specific components of the plant may be modeled to provide additional insight into these elements where modeling the whole plant would be either too complex or too expensive. Whether the whole plant or a component is modeled, it is necessary to design the model and convert the loads and/or dynamics measured in the scale model, to expected loads and/or dynamics of the full-scale system based on an appropriate scaling law. The appropriate scaling law is defined by the relative importance of the types of forces in the system. Except in very particular cases, the three important types of force that act on an FPV plant are inertial forces, gravitational forces, and viscous forces.

The two main scaling-laws associated with FPV plants are based on the Froude number and Reynolds number. Froude scaling ensures that the ratio of the inertial and gravitational forces is maintained in the scale model. This is the most common scaling law used for wave tank testing because waves are defined by a balance between inertial and gravitational forces. Reynolds scaling ensures that the ratio of the inertial and viscous forces is maintained in the scale model. This is typically used in aerodynamic modeling because gravitational forces can normally be neglected. Unfortunately, it is not generally possible to satisfy both Froude scaling and Reynolds scaling in the same model. Thus, care needs to be taken in the design of the scale model and the interpretation of data from physical modeling to ensure that it provides an accurate indication of the expected loads and dynamics of the full-scale system.

Analysis of System Response

It is usual to start the analysis of a floating structure using a frequency-domain analysis. This analysis provides two very useful results; the system natural frequencies and the system RAO. As a frequency-domain analysis is fundamentally linear these results are often not suitable for detailed design (where nonlinearity may have a significant effect) but provide insight into how the expected loads and dynamics of the system may vary with frequency.

A natural frequency of a system is a frequency at which the body will oscillate in the absence of an exciting force. Each natural frequency will have a mode of oscillation, defined by the relative motion in each of the degrees of freedom in the system. Typically, a system will have multiple natural frequencies and generally, the greater the number of degrees of freedom the greater the number of natural frequencies. Natural frequencies are important because excitation at these frequencies can cause very large displacements that have the potential to damage or even destroy a system.

A more complete representation of the system response is provided by the RAO, which provides the frequency response of a system to excitation, where the amplitude of response has been normalized by the amplitude of excitation. An example of a RAO is shown in Fig. 5.14 for the heave response of a floating body to wave excitation. It is common to calculate the RAO for a range of damping values as the damping is often nonlinear and so cannot be easily approximated in a frequency-domain model. However, because damping can have a significant effect on the response it is useful to determine the sensitivity of the system response to damping to provide additional insight into the system response.

The RAOs can be used with the excitation spectra from the waves, wind, and currents to help identify potential areas of concern in the design, where the basic

FIG. 5.13 QUB wave tank at Portaferry.

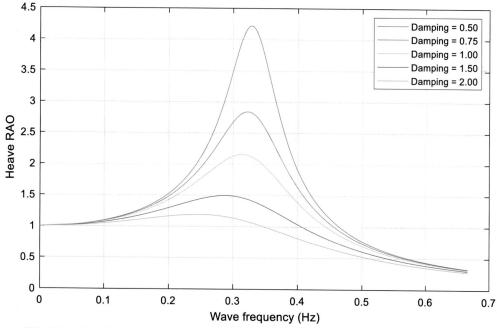

FIG. 5.14 Example of wave heave response amplitude operator for range of damping coefficients.

frequency response is calculated as the product of the excitation spectra and the RAO. If the excitation is small at the natural frequencies and the peaks of the RAOs, then this normally suggests that the response should be acceptable. The major advantage of this approach for the initial design of the structure is its relative computational simplicity and speed; major issues can be identified early, and alternative designs investigated with a relatively small amount of effort. This technique is especially useful because the RAO can more easily be linked to key characteristics of the design, which may be obscured in the construction and interpretation of a time-domain model.

In addition to first-order excitation, where the response is at the same frequency as the excitation frequency, low-frequency resonances can occur with coupling to second-order difference-frequency excitation, which is proportional to the square of the wave amplitude. The difference-frequency excitation can result in a problematic low-frequency response when the system has low-frequency resonance that is minimally damped. A common case where this may need to be considered is in the use of compliant moorings that have a natural period of response in surge/sway in the region of 40–120 seconds for large offshore structures, and the damping force is typically low because it is proportional to the velocity squared and

the low-frequency response velocities are typically small.

Another consideration associated with the resonant response of the system is vortex-induced vibrations and flutter. It has already been noted that constant flow past a slender element can result in an alternating transverse force due the asymmetric shedding of vortices (see Wind Loads section). This can be particularly problematic if the slender element has a natural frequency that is similar to the vortex-shedding frequency. The induced motion of the slender element is called "flutter." An important consequence of these vortex-induced motions is a potential increase in the loads, especially fatigue loads. Drag coefficients may also increase, which could affect the global analysis of the system.

Although significant progress in design can be made using a frequency-domain analysis of the system, a time-domain analysis is typically required to determine the fatigue and extreme loads. However, a time-domain analysis for a system with a large number of degrees of freedom (as would be typical for an FPV plant) is computationally expensive and so the number of simulations that can be completed is likely to be limited. A careful choice of a set of simulation conditions in terms of sea state and wind/current speed will generally be sufficient for generating data that can be used for a fatigue analysis; however, estimating the extreme loads is more complex.

Extreme loads are typically defined as the maximum loads that can be expected during a specific duration, e.g., the 50-year design loads. That is, the maximum loads that are expected to occur during a period of 50 years. Accurate prediction of the 50-year design loads requires consideration of two distinct aspects. The first aspect is the conditions that are likely to generate the maximum loads and their frequency of occurrence, while the second aspect is the accurate prediction of the maximum loads from the simulation data. Unfortunately, identification of the conditions that are likely to generate the maximum loads requires a full understanding of how the wind, waves, and currents affect the system dynamics and subsequently the loads. In particular, it is important to consider the constant wind or current loads that may offset the operating location so that snatch loads are then generated by the waves. Typically, more than one set of conditions are identified as potentially causing the maximum loads, and these need to be simulated and the resultant loads investigated.

Unless a suitable design code approximation can be identified, for each set of conditions, a time-domain simulation of a 3-hour event that contains approximately 1000 waves can be used to generate a suitable data set. A 3-hour simulation is typically considered to generate sufficient data to provide a good estimate

of the distribution of the extreme loads for the given conditions. These data are then used in an extreme value analysis, which processes these data to provide an estimate of the maximum loads that may be expected to occur for the specified return period. For example, Fig. 5.15 shows an extreme value analysis for a 3-hour simulation. If these conditions are expected to occur for a total of 30-hour within the specified return period, then the curve can be extrapolated to estimate the maximum load for this longer duration of these conditions. A key point to note in this graph, which is typical for extreme loads, is that the probability scale is logarithmic and so that a 10-fold increase in the duration of the simulation will only increase the expected maximum load by approximately 10%.

A particularly important aspect of the estimation of extreme loads is the case where there is a sudden increase in stiffness. There are two common situations where this may occur at the motion limit of a joint, which is often call the "end-stop" problem and when the mooring lines suddenly become taut, where it is considered to generate snatch loads. In both cases the fundamental issue is that the preceding compliance means that there is the potential for a significant amount of momentum to be generated. Effectively, reaching an end-stop or a mooring line limit of extension results in the requirement for a significant deceleration (assuming that the end-stop or mooring line does

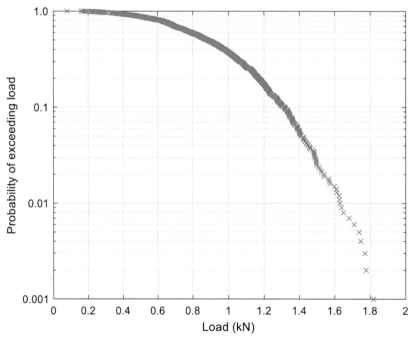

FIG. 5.15 Extreme value analysis.

not break) and the combination of significant momentum and rapid deceleration results in very large forces. Thus, where the potential for end-stop or snatch loads exist it is important that the modeling undertaken correctly represents the nonlinearity that will define these loads and thus provide accurate specifications for the structural and/or mooring requirements.

6. MOORING SYSTEMS

Introduction to Mooring Systems

A mooring or anchoring system is required to keep the FPV plant within an acceptable range of motion; this is called station-keeping. This range of motion should be such that the plant presents no threat of damage or hazard to itself or other users of the body of water, and this would normally be achieved by defining a maximum horizontal displacement. Examples of damage to itself would include collisions with the shore and seabed/lakebed, while hazards to other users would include occupying seaways and colliding with other watercraft. In addition, the mooring or anchoring system should be such that it does not allow the potential for damage to the plant's electrical connection.

A key requirement of station-keeping is the transmission of the loads to ground because without transmitting the loads to ground it is not possible to keep the plant in the right location when it is subjected to loads consistently acting in one direction. These loads will be generated across the whole of the FPV plant, although not necessarily with a consistent magnitude, and so the transmission of the loads to ground does not only concern the mooring system, but must also consider the transfer of the loads in the floating structure to the mooring system. Thus, the selection and design of the mooring system is closely linked to the design of the floating structure. At one extreme, it would be possible to include a mooring point for each float, which would minimize the requirement for load transfer in the floating structure but may be expected to increase the cost and complexity of the mooring system. At the other extreme, a single mooring point could be used to anchor the plant, but then this would increase the requirements for load transfer throughout the entire floating structure.

In addition to station-keeping, it is necessary that the moorings are designed so that they allow the structure to move vertically with variations in water level. In some circumstances, especially where the body of water is a reservoir that experiences significant changes in volume throughout the year, this can result in a requirement of several tens of meters vertical variation in the

position of the floats. In the case of plants with tracking, it is also important that the moorings are designed to allow the PV panels to track the sun where the movement originates in the floats supporting the PV panels. The joint requirement of station-keeping, together with allowing particular movements necessary for the effective operation of the FPV plant makes the design of the moorings particularly challenging.

The optimal combination of the number of mooring points and the floating structure design will be that which provides the lowest cost of energy solution. It is anticipated that this will depend on a large number of factors including the type of floating structure, the type of moorings, the plant size, the water depth, and the characteristics of the environmental loads. As there are a wide range of possible designs of FPV plants, it is not possible to provide guidance on an appropriate mooring design for every particular scenario. Moreover, it is anticipated that the optimal combination of mooring points and load transfer may change as the technology in FPV plant advances. Thus, the mooring and structural design for each proposal of FPV plant will need to be assessed based on its unique combination of situational and environmental factors.

Although the optimal design of moorings for FPV plants is still a long way from being resolved, the mooring of other structures can be used to provide an indication of the types of moorings that are available. These can be considered as fundamental building blocks or options around which a specific mooring design can be developed. This should not be considered to preclude other possibly novel solutions that may arise as a better understanding of the particular demands and costs for FPV plants is developed. Rather, these current solutions should be considered as potential starting points for actual mooring designs.

Rigid Mooring System

A completely rigid connection between the ground and a PV plant would mean that it is not floating; however, it is possible to have rigid elements that allow movement in heave, so that it is an FPV plant but limit motions in sway and surge to provide the required station-keeping characteristics. An example of a rigid mooring system is shown in Fig. 5.16 and can be commonly found in marinas to secure the jetties and walkways. In the design of a rigid mooring the heave motion could be achieved by allowing rotation of one or more connecting arms, or by translation along a vertical column. Both of these solutions are common in the mooring of floating pontoons, with connecting arms normally used when the range of vertical motion

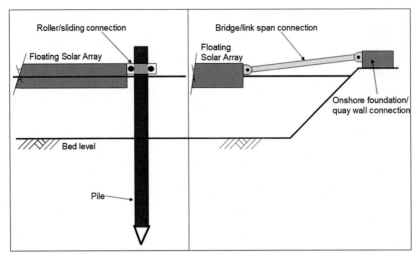

FIG. 5.16 Examples of rigid mooring systems.

is relatively small and translation along a vertical column normally used when there is the potential for large low-frequency vertical motions, e.g., variations with the tide.

In circumstances where the variation in the mean water level is relatively small then it may be possible to provide rigid moorings to the shore, which could also be used to provide access to the plant for operations and maintenance. However, except is the special case where there is a conveniently sized bay in the shoreline, a rigid connection to the shoreline is likely to experience high yaw moments due to the distance between the shore and an environmental force that has a line of action approximately parallel to the shoreline.

A rigid mooring to a point within the body of water can be achieved by the installation of one or more piles. A particularly interesting possibility is the installation of single piles, or groups of piles, around which the FPV array can rotate to allow tracking of the sun. However, without supplementary moorings, the use of piles is likely to be limited to relatively shallow water due to the large pitch/roll moments in deeper water due to the vertical distance between the water surface (where the loads will primarily act) to the bed (where the pile is anchored).

Fig. 5.17 shows an example of a floating solar array located centrally by a single pile held by a large concrete block on the bottom of the basin, which is 4 m deep. In this case, the plant has an installed capacity of 200 kWp and is equipped with a tracking system. The complete system rotates around the central pile driven by a small thruster.

Taut Mooring System

A taut mooring system consists of a number of cables that are usually held in tension using the excess buoyancy of the floats as shown in Fig. 5.18. The cables themselves are connected to the ground using piles or clump weights. The cables in taut moorings are typically aligned vertically so that there is some freedom for the floats to move in a horizontal plane, but with a limited amount of motion available in the vertical direction. To avoid excessive variations in cable tension due to changes in surface elevation (due to waves or tides) the water-plane area of the float is normally kept relatively small. The fundamental characteristics of taut moorings means that in their traditional configuration they are only likely to be suitable for circumstances where there is only a small amount of vertical variation

FIG. 5.17 Single pile mooring system.

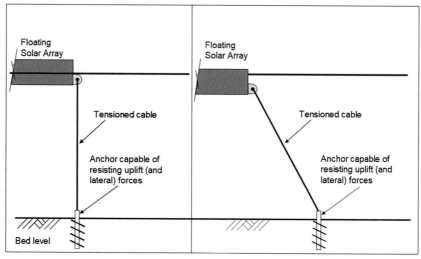

FIG. 5.18 Example of a taut mooring system.

in water depth. Moreover, the low cost of cable means that they could be particularly suitable for deployments in deep water.

Although taut moorings most commonly use vertical cables, other cable inclinations are also possible. By varying the inclination of taut moorings, it is possible to vary the relative constraint between the vertical and horizontal motions of the floats. By replacing the traditionally steel cables with modern fiber ropes, it is also possible to provide a larger amount of compliance, which could be exploited in the mooring design to avoid potentially problematic mooring resonant frequencies and reduce structural loads. A possibly interesting solution that could be applied in relatively small bodies of water would be to have virtually horizontal taut mooring lines that restrict motion in the horizontal plane but allow relatively large amounts of motion in the vertical direction. In this case, the tension would not be maintained by excess buoyancy, but by tension in opposing ropes/cables. It would also be possible to use a set of at least three inclined taut moorings to provide a rotation point about which floats could rotate to provide PV tracking with the sun.

Catenary Mooring System

A catenary mooring system uses the self-weight of the mooring line to provide a variable vertical and horizontal spring rate to the moored float; the term catenary refers to the shape that the mooring line adopts due to its self-weight as shown in Fig. 5.19. For this reason, catenary mooring lines are typically chains, with the size

of the links defining the amount of self-weight. The characteristics of catenary moorings are that the mooring stiffness will increase as the float moves away from the catenary mooring anchor point. Because catenary mooring will produce a horizontal load on the float, they are generally deployed with mooring lines extending in at least three different and opposing directions. The shape of a catenary mooring means that mooring lines generally lie along the ground for a distance before it connects with the anchor. Thus, the mooring load exerted on the anchor is essentially horizontal, which may influence the choice of the anchor to be used with these types of mooring system.

The required length of a catenary mooring line depends on the water depth and the required range of motion. A significant consideration in the design of catenary mooring systems is the rapid increase in the

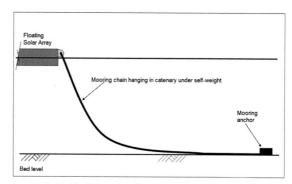

FIG. 5.19 Example of a catenary mooring system.

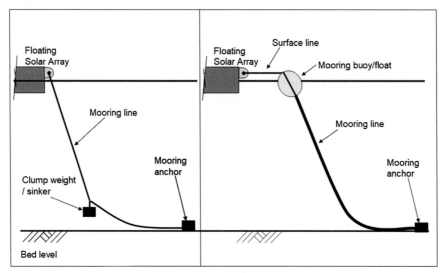

FIG. 5.20 Examples of compliant mooring systems.

mooring stiff as the mooring line becomes almost straight, which can result in significant snatch loads as discussed in Analysis of System Response section. Thus, the required range of motion is often dictated by the requirement to avoid snatch loads. In addition, although catenary mooring systems can be suitable for moorings where there are significant variations in water depth, care is required in their design. Specifically, the characteristics of a catenary mooring system can change significantly with changes in the water level and so it is clearly important that the different characteristics of the moorings are considered in the design process.

Compliant Moorings

A compliant mooring system consists of one or more floats and/or weights that are connected to the mooring cable/rope in between the FPV plant and the ground as shown in Fig. 5.20. Compliant mooring systems have many of the characteristics of catenary mooring systems but can generally be deployed with less space and create less disturbance to the bed of the water body as the mooring lines are not continually lifting, replacing, and moving on the bed. This could be a significant factor in the design of the mooring system where it is important for water quality or otherwise that the bed experiences minimal disturbance.

REFERENCES

[1] Open University, Waves, Tides and Shallow Water Processes (Chapter 1), second ed., published by Open University & Butterworth-Heinemann, 1999. ISBN 0750642815.
[2] Coastal Engineering Research Centre, Shore Protection Manual, fourth ed., US Army Corps of Engineers, 1984.
[3] S.K. Chakrabarti, Hydrodynamics of Offshore Structures, WIT Press, Southampton, UK, 1987.
[4] T. Sarpkaya, Wave Forces on Offshore Structures, Cambridge University Press, 2010. ISBN 9780521896252.
[5] R.I. Harris, The Nature of the Wind, the Modern Design of Wind Sensitive Structures, CIRIA Publications, London, 1971.
[6] C.S. Durst, Wind speeds over short periods of time, The Meteorological Magazine 89 (1960) 181–186.

FURTHER READING

[1] DNV, Environmental Conditions and Environmental Loads - Recommended Practice, DNVGL-RP-205, 2017.

CHAPTER 6

Cooling Systems

MARCO ROSA-CLOT • GIUSEPPE MARCO TINA

1. INTRODUCTION

It is well known that high photovoltaic (PV) cell temperatures cause not only a decrease in electrical efficiency for main commercial PV technologies but also a reduction of the lifespan of PV modules. The latter negative effect is due to increased thermal fatigue, causing excessive mechanical stress and so an increase of risks of cracks in the PV cells (Fig. 6.1).

PV silicon crystalline cells experience the highest efficiency drop with the rise in temperature, with a magnitude of approximately 0.45% °C. To mitigate such negative effects, in recent years, different cooling techniques have been proposed and tested experimentally [1−3]. The decrease of the PV cell temperatures depends on cooling techniques, type and size of the module, geographical position, and the season of the year. Several cooling techniques have been proposed based on the following techniques:

1. Forced air
2. Water veil
3. Water spraying
4. Circulation of forced water
5. Water submersion

Many reviews and comparative analysis on these techniques have been done, see Refs. [4,5], but this chapter focuses on water cooling techniques as they are surely the most cost-effective solutions for floating PV (FPV) plants.

Obviously, to adopt water-based solutions, two elements are critical: enough availability of water and low energy consumption for pumping, conditions that are satisfied in FPV plants.

2. EFFECT OF WATER ON PV CELLS

The most efficient way to obtain electrical energy is from direct solar irradiance via photovoltaic effect. In the last 10 years, the efficiency of average commercial wafer-based silicon modules increased from about 12% to 17% (Super-mono 21%), at the same time CdTe module efficiency increased from 9% to 16%.

The "standard test conditions" or STCs are universally applied to rate peak power output of a solar cell in a laboratory or a module out in the field, but rarely occur in real outdoor applications. Under STC protocol, solar modules are characterized in a controlled environment where the cell temperature is maintained at 25°C.

Generally, the irradiance is smaller and the temperature is higher. Both factors reduce the power that can be delivered by the module, so more realistic power output must be calculated, such as:

- Nominal operating thermal conditions which are based on 800 W/m^2 with 20°C ambient temperature and a 1 m/s breeze, of the panel. A reasonable value of the temperature in this case is 48°C (silicon solar cell).
- Photovoltaics for utility scale application (PVUSA) or more simply test conditions characterization procedure (PTC). In this last case, the module is kept 10 m above ground, while its ambient temperature is maintained at 20°C. Wind speed of 1 m/s and absolute air mass of 1.5 are also kept [6].

Efficiency Reduction due to the Thermal Drift

The presence of water on PV cells determines three main effects: reduction of operating temperature, solar spectrum modification, change in the reflection. In general, it has been shown that the overall efficiency of photovoltaic cells drops drastically with an increase in temperature. The manufacturers of PV modules provide in the PV module datasheets the maximum power thermal coefficient, kp, measured in % °C. The constant kp ranges widely from one technology to another as shown in Fig. 6.2, whereas the monocrystalline cells experience the highest values of P$_{max}$ reduction.

The impact of thermal losses on the yearly production of PV plants depends on many factors (e.g., PV technology, ventilation, climate). In Ref. [7], a comparative analysis of PV plants performances considering climate effect has been done; the thermal losses in two PV plants have been measured, specifically we

Floating PV Plants. https://doi.org/10.1016/B978-0-12-817061-8.00006-3

FIG. 6.1 Panel with cooling through water veil.

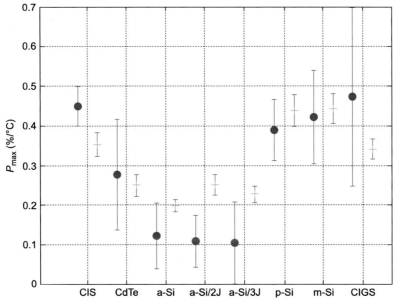

FIG. 6.2 Range of kp of modules declared by manufacturers [3].

have the following results: 3.9% in the north of Denmark (Aalborg) and 6.6% in the south of Italy (Agrigento). The thermal losses are referred to the cell temperature of 25°C, whereas if cooling techniques are applied a lower temperature can be reached so in turn a lower temperature can be reached with a larger increase of the yearly energy production.

In the case of concentrated PV cells, which use concentrated sunlight to produce larger amounts of power, and reduce the cost of generally expensive PV equipment, it has been observed that high temperatures greatly decrease the working life of the whole PV system.

Cooling mechanisms have already been proposed, see Refs. [1,8–11], and it has been shown that a sizable amount of power can be gained by the utilization of a cooling system.

Nevertheless, since a large amount of irradiated energy converts into heat, recent developments have been concentrated on harnessing that waste heat into useful thermal energy. Generally, hybrid elements that harness both electrical and thermal solar energy are called photovoltaic-thermal units (PV/T unit). These units usually have a higher overall efficiency but lower specific efficiencies, when compared with stand-alone photovoltaic and solar collectors.

Another important issue is that it is common practice in the market to sell or buy PV modules covered by a 20-year warranty. The warranty is supposed to cover safe operation (no electrical, thermal, mechanical, and fire hazards) and acceptable level of performance. This "acceptable level of performance" is commonly set at 1% degradation a year. Very often the limit of 1% degradation is reached and exceeded or, even worse, the panel breaks.

Solar Spectrum Modification

PV cells cannot convert all the incoming solar radiation into electricity but only a small amount. In fact, they convert the portion of solar radiation mainly in the visible spectrum, between 400 and 700 nm. The other well-known main components of the solar spectrum are the ultraviolet radiation and the infrared radiation. The infrared component is a large section of the solar spectrum that is absorbed by the PV cells but not converted into electrical energy and results in an increase of temperature.

The amount of short circuit current generated by a PV cell is proportional to the solar light intensity. Changing the spectral composition of light that strikes the cell surface will also influence the current output as the quantum yield of solar cells varies with the wavelength of light. So it is crucial to evaluate the effect of water on solar radiation spectrum. In physics, absorption of electromagnetic radiation is the way by which the energy of a photon is taken up by matter. Thus, the electromagnetic energy is transformed into other forms of energy, for example, into heat. Usually, the absorption of waves does not depend on their intensity (linear absorption), although in certain conditions (usually, in optics), the medium changes its transparency depending on the intensity of waves going through, and the saturable absorption (or nonlinear absorption) occurs.

Pure water is a strong light absorber, the absorption strongly depends on the wavelength [1], but fortunately this absorption occurs mainly in the red-infrared region. Particularly, the transmission of light in pure water is maximum in the interval of wavelength between 350 and 550 nm.

The gain or loss in efficiency of a PV panel in a standard position, placed outside the water, has been widely

discussed in Refs. [2,12] taking into account the solar spectrum dumping at a given water depth, folded with the efficiency spectrum of PV material, corrected by the temperature drift.

However, in the case of a few millimeter water veil the effect of water absorption is completely negligible, and it is not necessary to introduce any correction to the PV module response.

The Transmitted and Reflected Light

The input of incident angles is used to evaluate the percentage of reflected radiation. This is easily obtained by using the Fresnel formula and the refraction index of water: $n_w = 1.33$, of glass $n_g = 1.55$ [12].

We want to highlight three points:
- Water reflection is considerably reduced with respect to the glass reflection (2% vs. 4.5%) for small incidence angles
- Water glass reflection is always very small
- The total reflection with a water layer is sizeably reduced since water plays the role of a graded glass

The plot of Fig. 6.3 gives the total light reflected in two different setups:
- radiation is reflected by glass (as usually happens in normal PV panel);
- radiation is reflected by the layer water glass.

The last dotted curve shows the gain in energy arriving on the PV modules since water reduces the incoming impendence of radiation and behaves like a graded glass. The gain in the transmitted light ranges between 2% and 8% as the incident angle increases and is plotted with a green line in Fig. 6.3.

This is an effect which plays an important role at high latitudes. In these locations, cooling effect is less important due to the relative low solar radiation, but the minor impendence entrance for sunlight can contribute by 4% to the increase of energy harvesting [10].

3. WATER VEIL COOLING SYSTEM

Water veil cooling (WVC) is an efficient and simple way for increasing panel efficiency. It consists in generating a water veil on the panel surface using a pumping system. In this way, the PV module behaves like a submerged one. PV modules in this condition have been largely analyzed in Refs. [12–15].

The effects of reduction of thermal drift and of reducing the light reflection add up and, depending on the latitude and on the climate conditions, give a

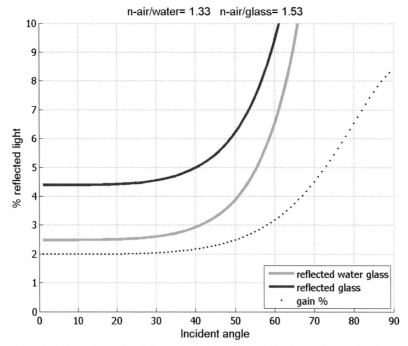

FIG. 6.3 Percentage of radiation reflected by glass and by the system water glass.

gain in energy harvesting which ranges from 8% to 10% in the temperate regions. In equatorial region, this value is expected to increase but strongly depends on the weather conditions. So, in arid zones, this value can reach 15%, whereas in very humid zones or when monsoon are active, this value is around 10%.

The negative effect of solar radiation absorption by water can be neglected due to the small depth of the water veil, which slightly affects the infrared part of the solar spectrum.

In Fig. 6.4, two P−V curves, with water veil (blue curve) or without (red curve), are reported: measurements have been taken on a PV module in Pisa. Radiation was 650 W/m^2 and panel temperature 45°C before cooling and 28°C when cooling was switched on.

This result shows that, even in cold months, water veil is effective and gives an important increase of the electric power. During sunny hot days, this power enhancement can reach 15%, and even more if concentrating systems (flat mirrors) are active.

For this reason, several efforts have been made to build plants which exploit this effect. However, it is practical only when working on a basin with the possibility to take water and to discharge it without any problem, like in FPV plants.

In the following sections, some details about a realized WVC are provided.

WVC Design Details

A WVC system is an irrigation system constituted by
a) a low pressure pump.
b) a set of polyethylene (or PVC) pipes positioned on top of each panel. Polyethylene pipes of 1.5″ having holes of 4 mm each 5 cm, and this allows a continuous water flow (see Figs. 6.5 and 6.6).
c) a control system for switching on the pump when the panel temperature exceeds a fixed threshold (30°C is a typical value).

The pump can be a normal commercial submerged pump used to empty the boat bilges.

Energy Balance and Gain

In order to save energy, the pumping system acts only on sunny days and when the PV temperature exceeds a given value (35°C) limiting the pump activity. In theory, we have to pump water at a height of less than 1 m, and using 1 kWh, we can pump more than 300 m^3 of water at this height.

Practically, we have to take into account losses in efficiency so that it is useful to take into account the

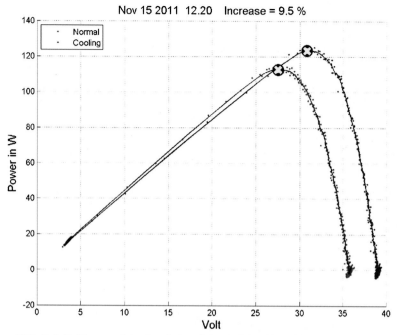

FIG. 6.4 P–V curve plots of a photovoltaic module with and without water veil.

FIG. 6.5 Rear view of the piping and cooling system.

parameters of a low-pressure pump as many are available on the market, and we see that with 1 kWh, it is possible to pump about 100 m³.

An estimate of a water veil on a single panel suggests that it is necessary to get about 2 L/min so that 100 m³ can supply a water veil for 1 hour for 100 PV panels. By the way, we remark that five pumps like this, with a very limited cost, are enough for a 1-Mp plant.

In conclusion, less than 1% of the produced energy is used for the pumping system, when active, whereas the gain in energy due to the better light absorption and to the absence of thermal drift is of the order of 10% or more.

FIG. 6.6 View of the water veil cooling system in operation.

It is worth to stress that the yearly energy gain is between 8% and 12%, but this is a temporal average along the whole year, whereas the pumping system acts only on the sunny days for a few hours around the middle of the day, and in this period the gain frequently exceeds 12%.

4. WATER SPRAY

A simple solution, alternative to the WVC system, is the use of high-pressure sprinklers. In this case, the cooling system consists of a few standard irrigation sprinklers which work at a pressure of 2−3 bar.

Several studies have investigated experimentally the performance of the PV cells with active cooling by water spraying. In Ref. [11], the impact of water spray cooling on the performance of the PV panel in highest solar irradiation level environment has been investigated experimentally [16].

In Ref. [11], a water spray cooling technique was proposed and experimentally tested on a monocrystalline PV module for different cooling options: cooling of front surface, cooling of rear surface, and cooling on both surfaces of the PV module. From one PV module test, on a short-term basis and without considering the energy consumption of the cooling process, the best cooling option turned out to be simultaneous cooling of front and rear PV module surfaces.

This simplified approach reduces the irrigation cost, but leaves unsolved the problem of the transparency of the water jet which can generate shadows on the PV module.

The results of the effect on PV modules cooling of the number and position of water jets have been reported in Ref. [11], where both sides of the PV modules were cooled at the same time by utilizing 20 nozzles, 10 on each side.

The results were measured for three different cases of cooling: front side cooling, rear side cooling, and both sides together, and were compared with noncooling case.

The research indicated that the water spray cooling has achieved a suitable effect on the PV panel performance and the best case was the simultaneous front and rear sides cooling PV panel (see Fig. 6.7). Lastly, depending on the experimental results, as presented in Table 6.1, the water spray cooling system had a proper impact on the PV panel performance. In Table 6.1, the effective increase of both power and efficiency include equivalent average power loss due to the water circulation system.

An experimental study about the impact of the water spray cooling on the performance of photovoltaic water pumping is reported in Ref. [17].

Three configurations are tested and compared: (A) two PV modules and 25 lit/h/module water spray; (B_1) three modules and 5 L/h/module water spray; and (B_2) three modules and 25 L/h/module water spray.

In cases A and B_1, the module temperature was decreased, and the reduction in case A was higher than case B_1 as shown in Fig. 6.8. The experimental results indicated that the system performance was significantly improved by spraying water on the PV module.

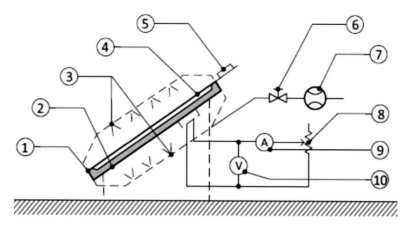

Legend:

1 – photovoltaic panel   6 – water flow regulating valve
2 – temperature sensor (back)   7 – water flow meter
3 – nozzles   8 – rheostat
4 – temperature sensor (front)   9 – ammeter
5 – pyranometer   10 – voltmeter

FIG. 6.7 Schematic layout of the specific experimental setup [11].

TABLE 6.1
Photovoltaic Panel Mean Performance Parameters for Different Cooling Circumstances [11].

Applied Cooling Options	Maximum Output (W)	Relative Increase Output (%)	Effective Increase Output (%)	Average Panel Temperature (°C)	Electrical Efficiency (%)	Increase in Electrical Efficiency (%)
Without cooling	35		–	56	13.92	
Back surface cooling	39.9	14.0	5.4	33.7	15.59	3.6
Front surface cooling	40.1	14.6	6	29.6	15.42	2.5
Simultaneous cooling	40.7	16.3	7.7	24.1	15.92	5.9

In Ref. [19], indoor tests were carried out by a solar simulator made of twenty 500 W halogen lamps, and four levels of irradiance were considered, in W/m^2: 413, 620, 821, and 1016. Two 50W monocrystalline PV modules were used in the test. A DC water pump was used to spray water to the front surface of one of the module and the other module was used as a reference. It was observed from the experimental results that the operating temperature of the PV module with water cooling system was reduced by 5–23°C, and the power output was increased by 9%–22%. So the water cooling is one way to enhance the electrical efficiency of the PV panel as shown in Figs. 6.9 and Figs. 6.10.

5. IMPACT OF OPERATING TEMPERATURE ON THE LIFETIME OF PV MODULES

The quality of silicon PV material is the primary determining factor of the conversion efficiency and initial cost for a solar module, where one gets what one pays

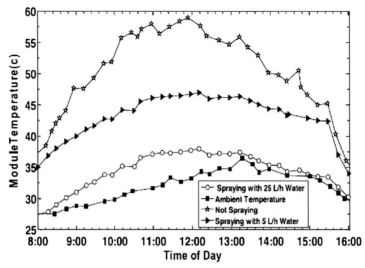

FIG. 6.8 Effect of water spray on the module temperature [18], cases A and B1.

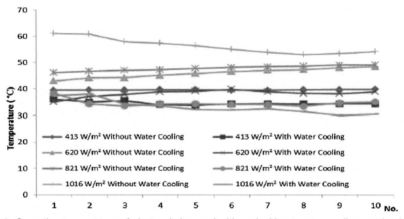

FIG. 6.9 Operating temperature of photovoltaic panel with and without water cooling mechanism [18].

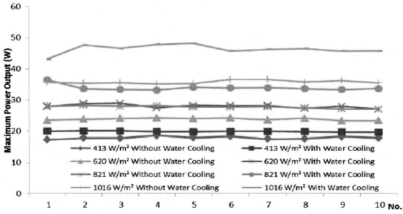

FIG. 6.10 Maximum power output of photovoltaic panel with and without water cooling mechanism [18].

for. Additionally, the PV modules have a double warranty, that is, product warranty with a range from 5 to 25 years (the most common value is 10 years) for manufacturing defects and performance warranty to insure a minimum efficiency after a given number of years of operation.

Today, for most of the modules, the guarantee on the yield is 90% of the nominal performances after 10 years and 80% after 20 years. Some manufacturers offer a multilevel performance guarantee, e.g., 95% for the first 5 years; 90% up to the 10th year; 87% up to the 15th year; 83% up to the 20th year; 80% up to the 25th year.

Such guarantees can be adversely affected by high operating temperatures that greatly increase the stress on solar cells. According to Ref. [15], the aging rate of a solar cell array doubles for every 10°C solar cell temperature increases.

To survive in harsh operating environments, PV modules have to be made of suitable materials that can keep the optical, chemical, electrical, and mechanical characteristics during the technical lifetime. To provide the requisite reliability, a PV module relies on packaging materials including protective superstrate, substrate, sealants, and encapsulants.

Several key properties associated with PV module reliability are critical for commercial success, these include (1) low-interface conductivity, (2) adequate adhesion of encapsulants to substrate, superstrate and PV cells, (3) low moisture permeation through all packaging materials, and (4) good mechanical properties such as tensile elongation and creep resistance at all operating conditions. Fig. 6.11 shows the correlations among the main environmental variables and the effects on the PV encapsulation materials as well as the failure of PV modules caused by packaging materials degradation under multiple stresses.

In Ref. [20], the authors investigate how properties change and/or degrade in polymeric materials used in PV modules and the correlation among materials degradation and failures of PV module and system performance in the field. It has been found that multiple stresses (IR radiation, temperature, and moisture) affect the adhesion strength between EVA/glass. In Ref. [22], the authors have found that higher temperature levels during exposure caused greater changes in the polymer morphology, which led to stiffening of the polymer. The significant increase in film stiffness under application characterized by relevant temperatures could cause some severe problems during the service lifetime of a PV module, starting with delamination and a reduced ability to withstand mismatches in thermal expansion and eventually cracking of the cell or the wiring.

In Ref. [17], a research about the effect of operating temperature on degradation of solder joints in crystalline silicon photovoltaic modules is reported. This study investigates the degradation of solder interconnections in c-Si PV modules for cell temperature rise from 25°C STC in steps of 1°C–120°C.

Solder degradation is measured giving the stress of the PV cell in Pascal for every °C of increase in the cell temperature respect to STC conditions. Three distinct degradation rates are observed.

- Region 1, with temperature between 25 and 42°C. This is the region where the stress manifests increasing from 0 to 10 Pa/°C.
- Region 2. In this region (42°C–63°C) the degradation rate reaches the 12 Pa/°C and cell damages are possible.
- Region 3, above 63°C is seldom reached in operative conditions.

Operations resulting in cell temperature between 43°C and 63°C are critical and induce maximum damage in the solder joint. This region is where the cooling mechanism should be more effective, reducing the damage of silicon cells.

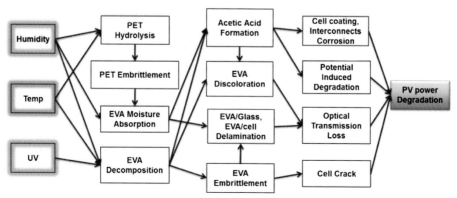

FIG. 6.11 Material caused by photovoltaic failure modes [20].

Finally, cracking in silicon solar cells is an important factor for the electrical power loss of PV modules. Aging effects due to environmental conditions (e.g., rapid temperature variations) are responsible for further propagation of crack, although it is nearly impossible to assess the individual contribution of each factor.

Analysis and statistics of degradation mechanisms in silicon modules observed in the field have reported various sources of failure of PV modules, namely, laminate internal electric circuit failure, glass breakage, junction box or cables failure, encapsulant decoloration or backsheet debonding, cell failures due to cracking. Among them, cell failure is considered to be responsible for 10% of the totally observed PV module failures, with an occurrence analogous to that of junction box or cables failure and to encapsulant decoloration or backsheet debonding.

6. CONCLUSIONS

The water cooling systems have evident advantages in increasing the efficiency and durability of a floating PV plant [2]. The advantages are the following: (1) a reduction of the thermal drift and as a consequences an increase of the PV module efficiency which in sunny days can reach the 10% or more, (2) a reduction of the reflectivity of PV modules which can rise the energy harvesting of 2%–3%, (3) last but not least a dramatic reduction of the thermal shocks which can increase considerably the cycle of life of the PV modules.

Some drawbacks are present. Apart for the maintenance problem quoted above, a cooling system is a dynamical mechanical apparatus and maintenance of the cooling system is necessary.

Furthermore we have to manage the freezing problem so that, during the winter period, when the temperature goes below zero, the piping network should be emptied.

Finally, the cooling system cannot be safely used in salty water or if the water is very dirty. In the latter case, a filtering procedure is needed, and it would be necessary to evaluate the costs of this operation. These problems can probably be solved, but experimental tests are necessary before drawing any final conclusion.

REFERENCES

[1] M. Rosa-Clot, P. Rosa-Clot, Support and Method for Increasing the Efficiency of Solar Cells by Immersion, 2008. PI2008A000088.

[2] M. Rosa-Clot, P. Rosa-Clot, G. Tina, Submerged photovolatic panel: SP2, Renewable Energy 35 (2010) 1862–1865.

[3] R. Cazzaniga, M. Cicu, M.-C. Rosa-Clot, C. Ventura, Floating photovoltaic plants: performance analysis and design solutions, Renewable and Sustainable Energy Reviews (2017).

[4] D. Sato, N. Yamada, Review of photovoltaic module cooling methods and performance evaluation of the radiative cooling method, Renewable and Sustainable Energy Reviews 104 (2019) 151–166.

[5] F. Grubišic Cabo, S. Nizetic, G.M. Tina, Photovoltaic panels: a review of the cooling techniques, Transactions of FAMENA (1) (2016) 63–74.

[6] R. Arndt, R. Puto, Basic Understanding of IEC Standard Testing for Photovoltaic Panels, TÜV SÜD Product Service, Peabody, MA, 2010.

[7] G. Tina, C. Ventura, D. Sera, S. Spataru, Comparative assessment of PV plant performance models considering climate effects, Electric Power Components and Systems 45 (13) (2017) 1381–1392.

[8] S. Krauter, Increased electrical yield via water flow over the front of photovoltaic panels, Solar Energy Materials and Solar Cells 82 (1/2) (2004) 131–137.

[9] S. Krauter, R. Hanitsch, Actual optical and thermal performance of PV-modules, Solar Energy Materials and Solar Cells 41/42 (1996) 557–574.

[10] R. Cazzaniga, M. Rosa-Clot, P. Rosa-Clot, G. Tina, Floating tracking cooling concentrating (FTCC) systems, in: 38th IEEE Photovoltaic Specialists Conference (PVSC), Austin (USA), 2012.

[11] S. Nizetic, D. Coko, A. Yadav, G.-C. Filip, Water spray cooling technique applied on a photovoltaic panel: the performance response, Energy Conversion and Management 108 (2016) 287–296.

[12] M. Rosa-Clot, P. Rosa-Clot, P. Scandura, G.M. Tina, Optical and thermal behaviour of submerged PV solar panel: SP2, Energy 32 (2011) 17–26.

[13] G. Tina, M. Rosa-Clot, P. Rosa-Clot, Electrical behaviour and optimization of panels and reflector of a photovoltaic floating plant, in: Proceedings of the 26th EU PVSEC. , Hamburg, Germany, 2011.

[14] Y. Ueda, T. Sakuray, S. Tatebe, A. Itoh, K. Kurokawa, Performance analysis of PV system on water, in: 23rd European Photovoltaic Solar Energy Conference and Exhibitio, Valentia, 2008.

[15] Y. Irwan, L.I.M. Fareq, H.I. Safwati, A. Amelia, Comparison of solar panel cooling system by using dc brushless fan and dc water, ScieTech Journal of Physics: Conference Series 622 (2015).

[16] H. Bahaidarah, A.G. Subhan, S. Rehman, Performance evaluation of a PV (photovoltaic) module by back surface water cooling for hot climatic conditions, Energy 59 (2013) 445–453.

[17] M. Abdolzadeh, M. Ameri and A. Mehrabian, "Effects of water spray over the photovoltaic modules on the performance of a photovoltaic water pumping system under different operating conditions," Energy Source Part A, vol. 33, 2010 pp. 1456-1555.

[18] A. Zubeer, H. Nohammed, M. Ilkan, A review of photovolatic cooling techniques, E3S Web of Conference 22 (2017) 200–205.

[19] Irwan, Indoor test performance of pv panel through water cooling method, Energia Procedia 79 (2015) 604–611.

[20] E. Wang, C. Peng, C. Tsai, I. Chou, C. Wang, Correlation between material degradation Behavior and PV module performance, in: International Photovoltaic Science and Engineering Conference, Busan Korea, 2015.

CHAPTER 7

Tracking Systems

MARCO ROSA-CLOT • GIUSEPPE MARCO TINA

1. INTRODUCTION

The idea that putting solar panels on water is not only possible but also economically and environmentally advantageous was coupled, since the beginning, with the concept that these kinds of installation should be rotating.

When the cost of solar panels was 10 times or more than the present cost, a large number of land installations were built with rotating systems, mono or multiaxial, but when the ratio $/W dropped down dramatically the percentage incidence cost of the supporting structure and its maintenance got much too high thus limiting the number of applications. More recently, however, horizontal axis tracking (HAT) has lowered its price and large plants have been installed in the equatorial region [1].

These solutions can take new life from the FPV technologies. Actually, on the water, both vertical axis tracking (VAT) and HAT are possible, and we can even find some examples of two axis tracking realizations [2].

VAT is easy and quite natural. It takes very little energy to move a floating object, and there is no need of heavy and costly mechanical support. See analyses of these systems done in Refs. [3,4].

HAT is also possible and will be done along the lines of recently installed large plants.

Without entering into the details of all these confined solutions, we will concentrate on the two most promising approaches:

- VAT: without confinement and with rotation generated by bow thrusters.
- HAT using fix platform based on the gable concept (see Chapter 4).

Advantages of energy harvesting for these solutions are analyzed in Section 2 of Chapter 3.

2. VERTICAL AXIS TRACKING

The idea of VAT for floating photovoltaic (FPV) plants was first shown in an artistic rendering which became very popular and inspired (erroneously) several practical solutions.

Confined Solutions

If we set a reference system like that in Fig. 7.1, we see that any position on the XY plane has the same potential energy. In a real situation, we can imagine that moving the floating object would be almost inexpensive in terms of energy requested. These obvious considerations convinced us of one important point, i.e., if we want to get the maximum benefit from our special condition, we should adopt a monoaxial VAT.

This approach was applied in our patent where "confined" VAT concept was further developed [5] (Fig. 7.2).

This patent clearly inspired several constructors. One of the more interesting solutions is given in Fig. 7.3, which seems to us a nice and interesting realization of our concept.

We can see the central portion (the payload) that carries the panels and that is able to rotate via the perimeter structure that is fixed to the bottom of the basin [6].

This kind of solution has a lot of benefits. The main one is that you can arrange a simple motor on the edge of the system able to obtain the rotation and, more important, you can stop it when needed. Braking the system is mandatory, and this geometry allows us to

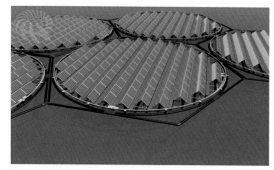

FIG. 7.1 Rendering of a floating photovoltaic with tracking system (with external confinement).

Floating PV Plants. https://doi.org/10.1016/B978-0-12-817061-8.00007-5

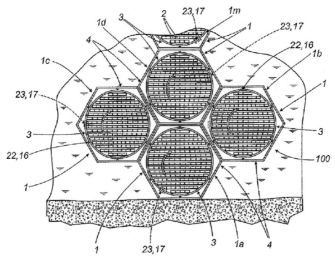

FIG. 7.2 Koiné Patent for tracking system [5].

FIG. 7.3 Floating structures with vertical axis tracking (Korea Water).

do it in a simple way since the scheme is similar to the brakes of a bicycle.

The reason why we did not do anything of this kind is simple: all kinds of perimeter structure that we could devise showed that their production cost is very high and the worst part is that the cost increases with the size more than proportionally.

The yield gain that we can obtain, thanks to the rotation, would never be enough to cope with the capital needed to build a system with a perimetral structure. It is true that this kind of solution can give very good precision relating to the angle of tracking, but this interesting feature is not necessary. These are not concentration systems where the accuracy needed is rather high. Actually, in VAT systems, an error of a 10 degrees does not change the total yield more than 1%, so tracking is useful but the solar pointing can be quite rough.

Other Solutions with Partial Confinement

Two other possible tracking systems were discussed in Refs. [3,4].

1. A system driven by winches like the one (Fig. 7.4) recently constructed by Sunfloat [7]. We have also

FIG. 7.4 Two axis tracking. The rope system which allows the vertical axis tracking is evident in this photo.

FIG. 7.5 Suvereto plant with central pole fixed on a concrete block on the basin bottom.

checked the solution with ropes on our Colignola plant [8], and we have verified that the practical managing of long ropes, winches and a fixed pole is quite complex and requires care and continuous management. The tracking is very precise but, notwithstanding the apparent simplicity, there are safety problems and the platform diameter should not exceed 50 m.

2. A wheel-hub system. This solution has not been realized but only studied and is suitable only for small plants. See Ref. [4] for further details.

3. VAT WITHOUT CONFINEMENT: BOW THRUSTERS

Looking for a solution of the tracking problem, we decided for a very simplified structure. As already mentioned, the first decision was to abandon the idea of any form of external containment of the system, but we still had to design something able to keep the whole structure in position and to allow the rotation around a center.

The first method experimented by the authors but suitable only for low depth basin is shown in Fig. 7.5. In this case, a big steel cylinder 8 m high with a diameter of 1.5 m is fixed on a large concrete block put on the basin bottom; depending on the water level, the emerging part

of the pole changes, but its position remains fixed and is the center of rotation of the tracking system. This solution is simple and very practical but works only for low depth since for high water levels, the forces and momenta acting on the pole foundation would be too strong.

In alternative, and not limited to shallow water, a system of chains connected to the middle of the rotating structure like the one in Fig. 7.5 is useful. Three concrete blocks at the corner of an equilateral triangle are chain-connected to a nozzle just behind the center of the platform pole (Fig. 7.6).

This system is well known and allows us to fix the rotation center approximately and can withstand wind load forces with a reaction force that increases linearly with the displacement (see Ref. [3]). This solution was adopted in the system with tracking and reflectors built in Colignola (Pisa) [8].

Once the center of rotation of the floating platform is blocked or anyway is stable with the possibility of moving by only few meters, the tracking mechanism can be implemented; the platform is moored to its center and one or two bow thrusters generate the couple for the rotations shown in the inset in Fig. 7.7. A windlass with an anchor enables us to fix the position when necessary. This solution is cheap and has been implemented and tested on our pilot plants both in Pisa and in Suvereto.

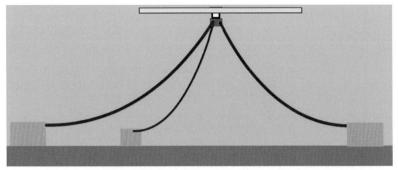

FIG. 7.6 Mooring system for a vertical axis tracking floating platform.

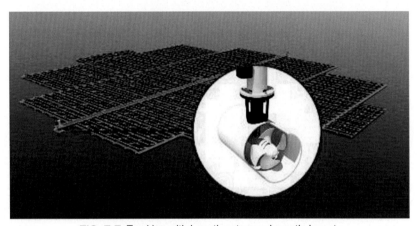

FIG. 7.7 Tracking with bow thrusters: schematic layout.

The full system is then composed of
- an electronic guidance system (EGS), which is able to recognize the sun position with respect to the platform: this is based on a camera and on a software able to identify the maximum radiation zone with a precision of a few tenths of one degree;
- one or more electric outboard motors (bow thrusters) positioned at the edge of the platform (only one in the Suvereto platform since the steel pole is a fix point of rotation, two in the Colignola plant in order to avoid too large a displacement of the center of the platform);
- a mooring post, limiting the rotation of the platform to a certain angle to avoid cable twisting when the system is stopped. By turning on and off the two motors, the EGS sets the platform in the correct direction toward the sun (Fig. 7.8).

Simulations and measurements of wind load and of structure strength have been carried out, and we have verified that strengths involved can be easily managed with a low power bow thruster.

What we need now is a system that can maintain the platform in position. A device that can provide a solution is a windlass like the ones used on boats to release or pull up the anchor. We can place the windlass in a position like the one showed in Fig. 7.9.

The end of the chain can be fixed onshore, if convenient, or on the bottom of the seabed hooked to an anchor or a concrete block. At night or when we want to turn off the system, we can just pull the chain up to a tension allowing only little movements. In this condition, the platform will face the South indefinitely.

Now the platform is expected to stay in place for some time, but there are external forces that can change the situation. During the day the system rotates following the sun, thanks to the bow thruster, and the platform will stay in whichever position it is placed if there is no wind

FIG. 7.8 Tracking system with bow thruster implemented on the Suvereto platform.

FIG. 7.9 Windlass anchoring system in Suvereto plant.

or a very weak one, but beyond a given threshold the platform will begin to rotate.

To restore the desired position, we can activate the motor but we must always balance the cost of the energy spent to push the system against the energy yield. A theoretical calculation of this balance is quite difficult so it was necessary to build a pilot plant to verify technical feasibility and economic convenience.

We realized two floating rotating plants: Colignola, where a reflector system was added requiring a much more precise pointing system, and Suvereto where a few degrees precision in the tracking are sufficient.

The Pointing System

Usually, in land-based photovoltaic tracking systems, the operating logic does not try to locate the sun really. Although the purpose is to point the system toward the sun, it is not necessary to have a camera that actually "sees" where the sun is placed in the sky. Let us suppose that coordinates on earth are exactly known. Then it is possible to refer to the ephemerides to define azimuth and elevation of the sun at any moment. This is the logic used by all the tracking systems because it is simple and efficient provided that you have a computer available.

Everything changes if the center of FPV is not fixed so that the position of the platform is not known. Then it is useful to disregard coordinates and focus on the azimuth of the sun using a camera [9]. Once we have a picture of the sky, we can elaborate the pixels finding the area where there are the pixels with the maximum brightness. Some other calculations can give us the angle that we must turn to orient the panels in the best position [10].

A system following these principles has been actually built. We developed also some electronics to couple it with a motor control device that gave some good results. In clear sky conditions and even with some clouds, we can follow the sun according to the specifications we have set because even with white clouds the system is able to recognize the position of the sun. In fact, we can operate as long as the human eye is able to locate the position of the sun in the sky and even further but when dark clouds cover part of the sky or worse, the system cannot operate. We can add that under these conditions the irradiation is low and tracking the sun is quite meaningless.

The tracking in all these cases cannot be based uniquely on a geometrical algorithm (astronomical tracking) since an exact fixed reference system is in general missing. The platform may move and the rotation center itself may be shifted by several meters due to the mooring system [11]. The adopted and field-tested tracking consists in a commercial camera positioned on the platform and able to take a wide-angle image of the sky. The acquired images are analyzed by a suitable SW which identifies the sunlight circle. In the absence of the solar disk, because hidden by cloud or fog, the system will target the area of the sky with the brighter zone. The precision of this apparatus is related to the camera pixel numbers. For a low-cost commercial camera, the resolution is below 0.5 degrees with an error which decreases in clear sky conditions. A more complex problem is the ability of our system to achieve the correct positioning of the platform. Our tests confirm that with the bow thruster's technique it is possible to follow the sun with an error of less than 2 degrees. The information so obtained is used to correct the platform position. Due to the low concentration system with the flat reflectors used there, the precision of 2 degrees was sufficient for the correct managing of the plant.

The small dimension of the platform in Colignola, however, was a limit since, due to the low inertia, frequent corrections were necessary and the bow thrusters were active for 5%–8% of the tracking time period. The use of this system on large platforms such as Suvereto is much simpler and more efficient.

Finally, we want to stress the importance of a safety mooring system during the tracking operation. This was implemented both in the Pisa and the Suvereto platforms using a simple winch with sinker and a short chain. Camera, bow thruster switches, managing of cooling system, and safety anchoring are all part of the plant control system.

Test of Tracking System

The tracking system on floating structures has been implemented by several companies facing the problem of orientation and of braking of the platform.

These problems have been analyzed and solutions have been implemented on our laboratory test in Pisa and on our platform in Suvereto and solved as follows [4]:

1. The tracking motion: In both platforms, we have used bow thruster. In Pisa, with a central chain mooring a couple of two 600 Watt electric outboard engines were used, whereas in Suvereto, thanks to the central rigid pivot, only one electric motor of 8 kW was sufficient (Fig. 7.12).
2. The pointing of the high-brightness area (not necessarily coinciding with the sun in the cloudy days): In both platforms the solution has been the use of a low-cost large-angle camera coupled to an image processing system.
3. Suitable software for managing the electric motors and the safety systems.

We want to stress that the tracking system is allowed to operate only when wind speed is below 10 m/s.

We collected several months of data about wind speed and we verified that the wind speed near the basin surface is strongly reduced (approximately by 30%) in respect of the wind intensity measurements at a 10 m quota, registered by local meteorological stations. Notwithstanding this reduction, we do not consider useful to maintain the tracking active when the wind speed is above 10 m/s. In this case, the system has to be blocked with panels south oriented, and this can be done by a suitable winch and anchor system.

If the wind is below this threshold value, the bow thrusters are switched every 20 minutes in order to position the platform correctly. If the wind moves the platform by more than 5 degrees, the bow thrusters are switched on in order to maintain the position.

In Ref. [4], there are some results of the web camera acquisition and of the corresponding computer analysis. The web camera takes a picture every 2 seconds and, even with cloudy sky, is able to find the maximum luminosity direction.

It was evident that even in the presence of a very cloudy sky the camera is able to identify the maximum of sky luminosity and to give the necessary coordinates to the board computer. Then a command is sent to the bow thrusters or, if the wind exceeds the threshold value, to the safety system which blocks the platform.

The results are positive, and the energy necessary to move the Suvereto plant never exceeds 2 kWh/day to be compared with the energy gain due to the tracking which can be more than 200 kWh/day.

We further remark that the platform dimension (and inertia) plays an important role and that we found a more regular behavior in the Suvereto platform (198 kWp) in respect of the small (30 kWp) Pisa platform.

Data for the Suvereto platform were collected from ENEL meter and refer to the south-oriented platform without tracking. The average for year 2012 and 2013 is given in blue with yellow dots in Fig. 7.10.

Systematic measurements were taken during the year 2014, and they confirmed the gain of 24.1% comparable with that predicted by simulation using the european Photovoltaic Geographical Information System (PVGIS).

4. THE HORIZONTAL AXIS TRACKING

As analyzed in Chapter 3, HAT can give relevant advantages especially for low latitudes. When we are within the tropical region, the gain in energy harvesting with respect to an optimal tilted fix plant ranges between 21% and 32%. In the temperate region, these values go down to 15%–25%, and the result is worst for high latitudes.

The problem that immediately emerges is the shadowing which for a land-based plant can be easily solved by increasing the occupied surface.

In Fig. 7.11, we can see the simple structure with the tracking system acting on a long strip of PV modules and the large space between two lines of tracking system.

Very large plants of this kind have been installed in the last 2 years, and in Fig. 7.12, a piece of the Mexico plant installed by ENEL is shown; see Ref. [13].

In Fig. 7.13, details are shown which highlight the need of very large spacing between PV rows in order to avoid shadowing effects.

Getting back to the water, a specific solution must be found since the raft has a cost which is essentially proportional to the occupied surface.

A single raft can support the tracking mechanism, quite similar to that in Fig. 7.12. In this case, a gable-like structure has been built with a rather large angle of the two equal inclined sides (in figure, 45 degrees),

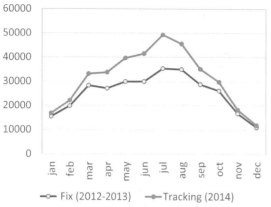

FIG. 7.10 Data from Suvereto plant.

FIG. 7.11 Green Source technology [12].

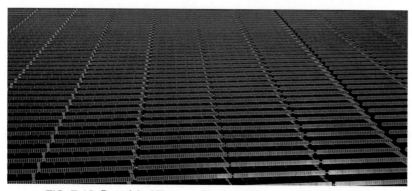

FIG. 7.12 Part of the Villeneuve (Mexico) EGP plant of 828 MWp [14].

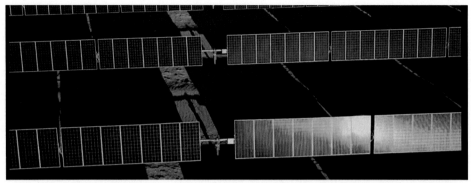

FIG. 7.13 Detail of EGP plant showing the large space between the rows of photovoltaic modules.

FIG. 7.14 Example of photovoltaic module positioning on the gable structure during the day.

see Fig. 7.14. This is necessary in order to allow a large east-west orientation of the PV modules.

The PV modules are aligned as in the gable structure, and each raft can hold a long central axis 12 m long coupled to a motor which allows a slow 90 degrees rotation during the day: 24 PV modules are positioned on a frame fixed to the axis and follow the east-west sun movement during the day.

The problem of wind load is not negligible, but during windy days the system could be positioned with PV modules in a horizontal configuration.

However, using this structure, the shadow effects would be very important and the energy harvesting strongly reduced. Then we cannot adopt the solution suggested in Chapter 4 for the gable structure: in this case, the rafts are adjacent and every single pipe is shared by two contiguous rafts. Here we need an adequate interspace.

The rafts are still connected but separated from each other by steel cables which are tensioned by the mooring system which should act on parallel lines of rafts. See Fig. 7.15, where the cables between rafts have been positioned just below the water surface in

such a way to allow the movement of a barge with a low keel for maintenance.

The structure is essentially the same as the gable structure, but we need to double the number of pipes. Due to the higher pipe number, the buoyancy is increased and so the dimension of the pipes can be reduced from 500 to 350 mm limiting the high-density polyethylene cost increase.

Using 3- to 4-m-long cables, the shadowing effect is strongly reduced at a reasonable cost but increasing the occupied area.

Due to the availability of water surfaces, especially in hydroelectric basins, this should be not a problem, and the great advantage of this structure is that this platform is fixed and does not move as in the case of VAT system. So the dimension of the plant is not limited and very large plants with HAT can be constructed.

The proposed structure leaves a wide space under the PV modules that could be exploited by installing flat aluminum reflectors (see Fig. 7.16) which should bring the very low water albedo (5%) to very high values (more than 50%) depending on the choice of

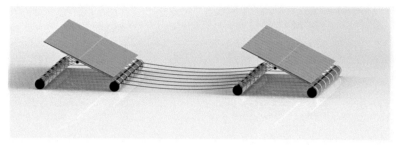

FIG. 7.15 Two rafts with tracking system connected by a steel cable.

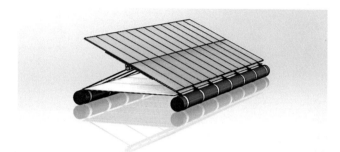

FIG. 7.16 Horizontal axis tracking system with a gable structure and flat horizontal reflectors.

materials and on the positioning and cleaning of this surface.

In our experience with the Colignola basin, we found that at a cost of less than 10 $/m², it was possible to get aluminum sheets with a reflection efficiency of about 90%.

This design, if coupled with bifacial PV modules, could raise the energy harvesting compared with a standard fixed land-based plant, remarkably.

REFERENCES

[1] M. Yaneva, P. Tisheva, T. Tsanova, The big Mexico Renewable Energy Report, in: MIREC Week 2018, February 2018.

[2] Floating Solar, Floating Solar Panels from Floating Solar, [Online]. Available: https://floatingsolar.nl/en/.

[3] M. Rosa-Clot, G.M. Tina, Submerged and Floating Photovoltaic Systems, Modelling, Design, Case Studies, Elsevier, Academic Press, London, 2017.

[4] R. Cazzaniga, M. Cicu, M. Rosa-Clot, P. Rosa-Clot, C. Ventura, Floating Photovoltaic plants: performance analysis and design solutions, Renewable and Sustainable Energy Reviews (2017).

[5] R. Cazzaniga, M. Rosa-Clot, P. Rosa-Clot, Apparatus for Tracking System, 2013. WO 2012 B51357 PCT 22-03-2012, Bologna.

[6] Y. Choia, N. Leea, A. Leea, K. Kimb, A study on major design elements of tracking-type floating photovoltaic systems, International Journal of Smart Grid and Clean Energy 3 (1) (2014) 70–74.

[7] Floating Solar, [Online]. Available: https://floatingsolar.nl/.

[8] R. Cazzaniga, M. Rosa-Clot, P. Rosa-Clot, M. Tina, Floating tracking cooling concentrating (FTCC) systems, in: 38th IEEE Photovoltaic Specialists Conference (PVSC), 2012. Austin (USA)FTCC.

[9] J. Yoo, Y. Kang, B. Song, J. Song, Solar tracking system experimental verification based on GPS and vision sensor fusion, Journal of Automation and Control Engineering 2 (2014) 417–421.

[10] Y. Choy, Y. Lee, A study on development of rotary structure for tracking type floating photovoltaic system, International Journal of Precision Enegeneering and Manufacturing 15 (11) (2015) 2453–2460.

[11] G.M. Tina, F. Arcidiacono, A. Gagliano, Intelligent sun-tracking system based on multiple photodiode sensors for maximisation of photovoltaic energy production, Mathematics and Computers in Simulation 91 (2013) 16–28.

[12] Green Source, [Online]. Available: http://www.greensource.com.tw/us/satrackers.

[13] T. Buckley, S. Kashish, Solar is Driving a Global Shift in Electricity Markets, IEEFA, 2018.

[14] Enel-Green-Power. [Online]. Available: https://www.enelgreenpower.com/stories/a/2018/05/villanueva-where-solar-energy-meets-the-digital-world.

CHAPTER 8

Integration of PV Floating With Hydroelectric Power Plants (HPPs)

MARCO ROSA-CLOT • GIUSEPPE MARCO TINA

1. HYDROELECTRIC POWER PLANT PENETRATION

Hydroelectric power plants (HPPs) represent by far the most important component of the Renwable Energy System (RES). At the end of 2017, they covered 16.4% of the worldwide electric energy production with increasing investments especially in China and in the equatorial regions. The HPP contribution reached 4185 GWh, and the total installed power was 1270 GW in 2017.

So, the contribution of HPP to the electricity production is overwhelming, and this is due also to the higher capacity factor (CF) which is given in Fig. 1.3 (Chapter 1) where it is compared with wind and solar energy production [1,2].

The installed distribution of HPP is given for macroareas in Table 8.1, which shows that East Asia with South America (mainly Brazil) has been the most important investor in the last few years.

However, the HPP installed each year is decreasing as shown in Fig. 8.1, and this seems to be true also for 2018 and for the forthcoming years. The minor investment in this sector is mainly from industrialized country, and in the last 5 years the reduction in investments has been of more than 10% per year and is supposed to continue.

Why this slowdown? There are several reasons:
1. The technology is mature. The first HPPs go back to the beginning of the 20th century so that many industrialized countries have almost reached the maximum exploitation of the location suitable for HPP.
2. The penetration of competitive RES like wind and solar. These technologies have a lower CF (2000 and 1000 hours, respectively, compared to the 3000 hours of the HPP) but can be installed in a very short time compared with the 10 years typically required by an HPP.
3. The environmental impact. This is by far the most important element in the slowdown of the HPP investments. Wind and solar plants have also an important visual impact, but HPPs require a more deep modification of the orographic system with consequences for human settlements.

The third point is without doubt the most important and has brought to a worldwide criticism from ecologists.

A simple illustration of what happens can be shown using the strongly criticized Balbina HPP as an example, which, however, is far from being the only one.

In Figs. 8.2 and 8.3, it is possible to see the effect of the Balbina Dam on the orographic structure of the Uatuma river (Northeast Brazil) where a 2.360 km^2 basin was created to generate an HPP of only 250 MWp [5].

In this case, the problem is related to the vastity of the zone flooded by water, to the displacement of many indigenous tribes, and to the emission of greenhouse gases due to the destruction of large area of tropical jungle.

However, other negative effects are present and historically documented, starting from the huge Aswan Dam on the high Nile (1960–70) which deeply modified Egypt agriculture, with the disappearance of the annual flood and heavy year-round irrigation, coast erosion, a saline intrusion, etc. and from many other examples in the tropical zone as well in industrialized countries [6].

So we can assume that new HPPs in the near future will take care of environmental impact and will carefully analyze the long-term consequences on the human settlements in the interested zones.

However, the floating photovoltaic (FPV) plants can be a strong support to the HPPS, since the two technologies can benefit from reciprocal advantages when working jointly.

Floating PV Plants. https://doi.org/10.1016/B978-0-12-817061-8.00008-7

TABLE 8.1
Installed Hydroelectric Power Plants: Power and Energy Produced in 2017 [3].

2017 Data	Power, GW	Energy, TWh	CF, hours
North and Central America	203	783	3857
South America	167	716	4287
Europe	249	559	2245
South and Central Asia	144	456	3167
East Asia and Pacific	468	1501	3207
Africa	35	131	3743
	1266	4146	3275

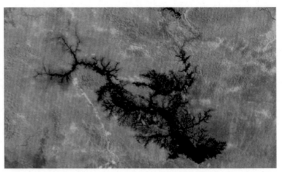

FIG. 8.2 Fractal structure of the basin generated by the Balbina Dam.

2. ADVANTAGES OF COUPLING FPV AND HPP

Several factors suggest the advantages of coupling photovoltaic (PV) plants with hydroelectric power stations, as analyzed by several authors [7–9]. In short, we can summarize the benefits gained by a hybrid PV-Hydro coupling as follows.

1. **Grid connection.** Artificial hydroelectric basins are equipped with power generators and are grid connected, so it is possible to exploit the existing infrastructure reducing the cost of a FPV plant.
2. **Reduction of power fluctuation.** In temperate regions, Italy, for example, the PV panels give the maximum energy yield during the hot season when the HPPs register a reduction of power due to the seasonal water cycle. This partial anticorrelation allows a not negligible reduction in the yearly fluctuations of electric energy production.
3. **No land occupancy.** The main advantage of FPV or submerged PV plants is that they do not take up any

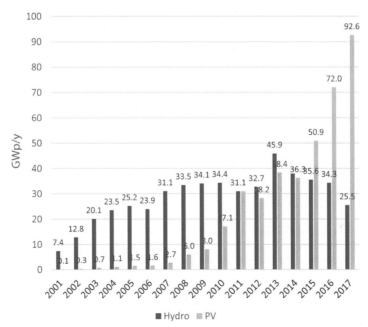

FIG. 8.1 Worldwide yearly hydroelectric power plants and photovoltaics (PV) installed in GW [4].

FIG. 8.3 Balbina Dam in Brazil generates 2360 km^2 shallow water basin.

land, except the limited surfaces necessary for electric cabinets, and this is important if we want to couple PV to hydroelectric plants [10]. Actually, the orographic structure of the basins generated by HPP dams is quite complex, and in many cases it is not possible to find near free land for vast PV plants.

4. **Installation and managing.** FPV plants are more compact than land-based plants, their management is simpler and their construction and managing (including decommissioning) straightforward. The main point is that no fixed structures exist, and the mooring of floating systems can be carried out in a totally reversible way, unlike the foundations used for a land-based plant. This is a keen point since the installation of an FPV has to be done without any impact on dams and HPP structures, and furthermore we want enough flexibility to take into account quick water level variations.

5. **Water saving.** The partial coverage of basins has additional benefits such as the reduction of water evaporation. This result depends on climate conditions and on the percentage of the covered surface but in arid climates the water evaporation reaches 2 m per year. The reduction of the evaporation rate increases also the water availability for HPP generation power. This effect is small but positive.

3. HYBRID FPV-HPP AND GEOGRAPHIC POTENTIAL

Hydroelectric power is mainly based on the construction of dams and on the creation of large basins. Run-of-the-river (ROR) systems cover only a small part of hydroelectric production and are important mainly in the mini- and microinstallations. In general, most hydroelectric power comes from the potential energy of artificial water reservoirs created by a dam, which in several cases are equipped with a pumping system for facing high peak energy demands [11].

Power and Energy Density

We want to compare the power and energy density for HPP and FPV. To do this, we define the quantities $\rho_{P,H}$ and $\rho_{E,H}$ (power density and energy density for hydropower plants) and correspondingly the same densities for FPV, specifically $\rho_{P,FPV}$ and $\rho_{E,FPV}$

- $\rho_{P,H} = P_H/S_B$, where P_H in GW is the peak power of the hydroelectric plant and S_B in km^2 is the basin surface,
- $\rho_{E,H} = E_H/S_B$, where E_H is the yearly energy production in GWh.

These two types of density are not common quantities and are not usually given in the literature: their average values for the most important HPP in the world are 4.0 W/m^2 and 19.2 kWh/m^2, respectively.

The ratio of the two densities allows us to calculate the capacity factor $CF_H = E_H/P_H$ (hours) which for HPP has values ranging from 2000 to 5000, depending on the basin characteristics, with an average value of 4789 hours.

The corresponding quantities for FPV are $\rho_{P,FPV}$ and $\rho_{E,FPV}$ which depend on the geometrical structure of the FPV plant and on the local radiation. The capacity factor CF_{PV} ranges typically from 800 to 1800 hours depending mainly on the latitude and on the mean weather conditions.

The quantity $\rho_{E,FPV}$ depends also on the geometry of the plant (panels pitch and tilt) and has been discussed in Ref. [12] with the result that, for an optimal floating structure, the solar energy produced by modules in horizontal position should be increased for modules with optimal tilt and pitch. In the following, data taken from PVGIS, or alternatively NASA database, and elaborated with the PVsyst will be used.

Summarizing this discussion, we define a typical FPV plant where 16 PV modules are installed on a raft of 48 m^2 (Length 12 m × Width 4 m), and we assume that the power of each module is 0.36 kWp (for PV modules 2 × 1 m). So the power available on a 1-km^2 floating plant would be:

$$\rho_{P,FPV} = 0.36 \text{ kW}^*16/48 \text{ m}^2 = 0.120 \text{ kW/m}^2$$
$$= 120 \text{ MWp/km}^2 \text{ or W/m}^2$$

This value can change due to an increase of the power density of the PV modules (recently modules of the same surface 2 × 1 m^2 have been proposed with a power of 0.44 kW) and to a change in the raft geometry (see Ref. [13] where a "gable" solution, suitable for equatorial regions, allows us to put 24 PV modules on a double raft of surface):

$$\rho_{P,FPV} = 0.44*24/48 = 0.220 \text{ kW/m}^2$$
$$= 220 \text{ MWp/km}^2 \text{ or W/m}^2$$

This possibility is very attractive since the compactness of the power PV plant is a positive factor. However, in the following, we will use the more conservative value of $\rho_{P,FPV}$ equal to 120 MWp/km^2, much higher (a factor 30) if compared with the HPP value.

In order to calculate the $\rho_{E,FPV}$ value, we have to know the solar energy yield for the different locations which can range from 800 to 1800 kWh/y per kWp so that we find different results, but on average we find 149 kWh/y/m^2, a value about eight times larger than the corresponding $\rho_{E,H} = 19.2$ kWh/y/m^2. The reduction from factor 30 to 8 between power and energy density is of course due to the much higher CF for the hydroelectric plants.

Optimization

Since the factor $\rho_{E,FPV}$ is much larger than $\rho_{E,H}$, it is quite evident that the coverage with FPV of part of the hydroelectric basins can strongly improve the energy production. Advantages, as said above, are remarkable since grid connections and infrastructures already exist. But the most important issue is that installation cost and timing are reduced, thanks to the previous management of the flooded area.

From Ref. [14], we can learn how profitable the coupling of HPP and FPV is. The authors present an

TABLE 8.2
Relevant Parameters for the Longyangxia Hydro−PV Power Plant [14].

Hydropower Reservoir Normal Pool Level	2600 m
Minimum outflow of a hydro unit	50 m^3/s
Maximum outflow of a hydro unit	292 m^3/s
Average hydraulic head	100 m
Installed capacity	1280 MW
Average annual energy production	5940 GWh
Capacity factor	4640 hours
Basin surface	300 km^2
PV array installed capacity (land based)	850 MW
Average annual energy production	1494 GWh
CF$_{PV}$ factor	1756 hours
Occupied area	20.4 km^2

analysis of the Longyangxia hydro−PV power plant. This plant is constituted by a large HPP (power 1280 MW) and by a large land-based PV plant (850 MWp). See Fig. 8.4. Main parameters of the plant are given in Table 8.2.

If we summarize data using our parameters, we can say that $\rho_{E,H} = 5940/300 = 19.8$ kWh/m^2/y and that the PV plant occupies a surface which is 6.7% of the basin surface. Of course, for an equivalent FPV plant the surface would be much less and should not exceed 8 km^2, that is, 2.7% of the flooded area.

FIG. 8.4 Google Earth map of the Longyangxia basin with the photovoltaic power plant.

The managing of this hybrid power station is very interesting. The power produced by the PV plant is fully used, and the power of HPP is gently tuned to match the grid requirement. This can be easily done since the power peak and the energy production are considerably less than those of the HPP. Following reference [14], we remark that the hydroelectric power of the HPP can be tuned in a coarse way and can follow the variation of the PV plant. In this way, the total output can match the grid requirements and the use of water resources is proportionally reduced.

In the specific case of the Longyangxia plant, in Fig. 8.5, we illustrate the behavior of the two components of energy production during a sunny summer day when the PV production reaches its maximum. In this case, the PV produces 6680 MWh to be compared with the HPP production of 24,350 MWh. So PV supplies 21.5% of the total energy production of 31.030 MWh.

This result should be averaged along the year, considering also winter and cloudy day, but the global result is that the available power of the HPP is increased by about 20% without any changes in the infrastructures.

The question now is what is the optimum size for the coupling of a PV plant (possibly an FPV) with a large HPP? The target is to maximize the energy production without changes in the basin orography and in the existing infrastructures. So, in order to minimize costs and exploit the advantages of this hybrid system as far as possible, we suggest installing an FPV able to supply, at its best, the power of the HPP itself.

In the typical situation for a Mediterranean latitude (Catania, for example) the daily average energy yield can be calculated for different months assuming that for 1 MW HPP an equivalent FPV of 1 MWp has been installed: averaging on the different months we find that the contribution of the FPV for the month of July, April, and January is 33%, 26%, and 17%, respectively, with an average yearly contribution of 27%.

However, this result underestimates the FPV contribution. In fact, the large basins quoted in Table 8.1 have a CF value of 4896 hours, whereas the CF factor is much lower for Sicily HPP (approximately 2500 hours) so that the FPV contribution is actually much more important.

DATA FOR THE FIRST 20 LARGEST HPPS

Data for the first 20 largest HPP in the world are shown in Table 8.3.

After the basic information about power in GW and energy production per year in TWh, we give the CF value (in hours), the flooded area S_B, and the power for km², defined above as $\rho_{P,H}$. We remark that the value found for CF is larger than the CF quoted in Table 8.1. This is due to the greater inertia and storage capacity of the very

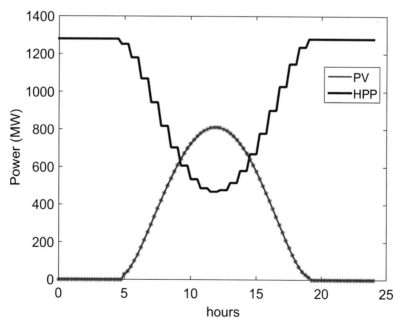

FIG. 8.5 Hydroelectric power plant (HPP) and photovoltaic (PV) power in MW during a sunny summer day for Longyangxia plant.

TABLE 8.3
Data for the Largest 20 HPPs in the World.

#	Name	HPP MAIN CHARACTERISTICS					FPV 10%				FPV OF POWER EQUAL TO HPP			Country
		Power, GW	Energy, TWh/y	CF, hours	Basin, km²	$\rho_{P,H}$, W/m²	Power, GWp	Energy, TWh/y	Area, km²	Area, %	Energy, TWh	Energy, %	CF Total, hours	
1	Three Gorges Dam	22.5	99	4391	1084	21	13.0	12.0	188	17.3%	20.7	21%	5312	China
2	Itaipu Dam	14.0	89	6357	1350	10	16.2	21.8	117	8.6%	18.8	21%	7700	Brazil
3	Guri	10.2	53	5218	4250	2	51.0	82.7	85	2.0%	16.6	31%	6839	Venezuela
4	Tucuruí	8.4	41	4950	3014	3	36.2	57.0	70	2.3%	13.2	32%	6527	Brazil
5	Grand Coulee	6.8	20	2937	324	21	3.9	4.7	57	17.5%	8.3	41%	4156	USA
6	Xiangjiaba	6.4	31	4761	96	67	1.1	1.0	54	56.2%	5.7	19%	5647	China
7	Sayano—Shushenskaya	6.4	27	4188	621	10	7.5	7.4	53	8.6%	6.3	24%	5174	Russia
8	Krasnoyarsk	6.0	15	2500	2000	3	24.0	23.2	50	2.5%	5.8	39%	3465	Russia
9	Nuozhadu	5.9	24	4085	320	18	3.8	5.1	49	15.2%	7.8	33%	5426	China
10	Robert-Bourassa	5.6	27	4719	2835	2	34.0	33.5	47	1.7%	5.5	21%	5702	Canada
11	Churchill Falls	5.4	35	6448	6988	1	83.9	80.3	45	0.6%	5.2	15%	7406	Canada
12	Bratsk	4.5	23	5006	5470	1	65.6	64.6	38	0.7%	4.4	20%	5990	Russia
13	Xiaowan Dam	4.2	19	4524	190	22	2.3	3.0	35	18.4%	5.6	29%	5854	China
14	Ust Ilimskaya	3.8	22	5651	1922	2	23.1	21.8	32	1.7%	3.6	17%	6597	Russia
15	Jirau	3.8	19	5093	258	15	3.1	4.6	31	12.1%	5.6	29%	6595	Brazil
16	Jinping-I	3.6	17	4722	83	44	1.0	1.3	30	36.4%	4.6	27%	5998	China
17	Santo Antonio	3.6	21	5922	490	7	5.9	9.1	30	6.1%	5.5	26%	7465	Brazil
18	Tarbela Dam	3.5	13	3738	250	14	3.0	4.3	29	11.6%	5.0	38%	5167	Pakistan
19	Ilha Solteira Dam	3.4	18	5197	1195	3	14.3	22.7	29	2.4%	5.5	31%	6783	Brazil
20	Ertan Dam	3.3	17	5152	101	33	1	1.7	28	27.2%	4.6	27%	6558	China
	Total/Average	**131.4**	**629**	**4789**	**32,840**	**4.0**	**394**	**461.8**	**1095**	**3.3%**	**163**	**27.0%**	**6029**	

CF, capacity factor; FPV, floating photovoltaic; HPP, hydroelectric power plant.

large basins and to the fact that small basins have a lower CF factor ranging from 2000 to 3000 hours.

We want to remark that this parameter can vary a lot from one plant to another but that the average value is quite low: 4.0 W/m^2.

The next two columns give the power of an FPV plant and its energy yearly production for km^2 under the hypothesis that 10% of the basin is covered by an FPV with $\rho_{P,FPV} = 120 \text{ W/m}^2$.

As evident, even limiting the coverage to 10%, the increase in energy production is sizable and in some cases it is more than the production of the HPP itself.

Using these values, we can see that the energy production, using a coverage of 10% of the basin, is 462 TWh, about 74% of that due to the hydroelectric power. This value is larger in equatorial zones where solar energy yield is larger, and smaller at high latitudes but anyway we can conclude that with a coverage of 10% the energy harvesting is substantially increased.

This result is very important, and for this reason the possibility of coupling FPV plants to HPP plants is quite natural and has been analyzed from the very beginning of the research in this sector [15].

The energy harvesting, however, is reduced and, averaging on the 20 PV plants, we get a CF_{PV} value of 1240 hours.

The other column of the table gives the results obtained by using the strategy suggested above, of installing an FPV power equal to the HPP power.

The first column of this block gives the area occupied by an FPV of equal power.

It is impressive to see that the surface occupied by the FPV plant (of equal power to HPP) is on average only a small fraction (3.3%) of the hydroelectric basin surface; actually this value changes a lot, from less than 1% to more than 50% and higher values are reached in structures where the out-of-river solution is favored for the HPP plants, but in most cases, the existence of large flooded areas justifies the solution of covering a small percent of it in order to double the RES power potential.

The other two columns give the energy produced and its percentage compared with the HPP energy yield, which ranges between 19% and 38%, with an average contribution of 27%.

This increases the CF (last column data) which rises from 4789 to 6029 hours.

We further remark that for small HPP plants with minor CF factor (in the range of 2000–3000 hours), the average contribution of the FPV plants should rise to about 45%.

4. A WORLDWIDE ANALYSIS

A similar but more detailed analysis can be done for the United States. We use the data published by Ref. [16] about 100 HPPs in the United States. In this case, the main results are synthetized in Table 8.4.

As evident, the high value of $\rho_{E,FPV}$ ($\rho_{E,FPV}/\rho_{E,H} = 38$) implies that by covering on average 2.5% of the hydroelectric basins, we can double the produced energy. However, we simply want to install FPV of power equal to that of the HPP. In this case, the surface of the basin occupied by FPV would be 1.19% of the full basin surface and the energy produced would be 40.5% of the hydroelectric energy production.

A more general analysis can be done using worldwide database for water resources. We refer to Ref. [17] for the free freshwater surface and to AQUASTAT database [18] for information about man-made reservoirs (MMR) and hydroelectric basin surfaces (HPPSs). See also Ref. [19] for the Digital Water Atlas database. The main results are collected in Table 8.5.

We start from the analysis done in Ref. [12]: the freshwater surfaces are taken from Ref. [17], and the data are collected for large geographic areas. MMR and HPPS are taken from AQUASTAT database [18].

A more careful analysis should be done and some minor corrections have been made to the AQUASTAT data [18]; however, the results are stable and can been synthesized as follow:

- The MMR represent 12.6% of the full water surfaces and the surface of HPPS represent 66%. This last percentage depends on orographic conformation, and a specific case is Japan where the presence of many basins at high quota and with small surface pushes this ratio below 10%.
- 380,948 km^2 of water basins originated by dams for HPP power plant are available around the world.

TABLE 8.4
Data for 100 US Hydroelectric Power Plants (HPPs) (Total Values and Average Data).

USA 100 HPP	Power, MW	Basin Surface, km^2	$\rho_{P,H}$, MW/km^2	$\rho_{E,H}$, GWh/km^2	$\rho_{P,FPV}$, MW/km^2	$\rho_{E,FPV}$, GWh/km^2
	32,574	22,736	1.43	5.02	120	191

TABLE 8.5
Worldwide Analysis of the FreshWater Reservoirs (FWRs), Man-Made Reservoir (MMRs), and Hydroelectric basin surfaces (HPPSs).

	FWR, km^2	MMR, km^2	HPPS, km^2	MMR/FWR, %	HPPS/MMR, %
Africa	540,030	46,499	24,197	8.6%	52%
America Central	58,801	4161	2899	7.1%	70%
America South	381,710	65,000	53,863	17.0%	83%
Asia South East	153,490	32,231	22,929	21.0%	71%
Asia South—India	48,320	1283	1081	2.7%	84%
Australia + New Zealand	58,920	4965	1216	8.4%	24%
Canada	891,163	97,914	95,224	11.0%	97%
China	270,550	12,979	7454	4.8%	57%
Europe (North)	178,156	30,267	24,724	17.0%	82%
Europe (South)	19,612	3091	2066	15.8%	67%
India	314,000	102,775	13,361	32.7%	13%
Japan	13,430	1394	130	10.4%	9%
Middle East	140,190	26,259	10,775	18.7%	41%
Russia	720,500	85,408	84,761	11.9%	99%
Turkestan	76,110	17,247	14,582	22.7%	85%
USA	685,924	43,904	21,686	6.4%	49%
Total	**4,550,906**	**575,377**	**380,948**	**12.6%**	**66.2%**

Now the question is how large is the FPV surface necessary to install an equivalent power? And how much is the energy produced? Table 8.6 answers these questions under the hypothesis discussed in Section 3.

It is remarkable that the surface necessary to install an FPV power equivalent to that of the HPP is only on average 2.38% of the basin surfaces and that in only three cases it is larger than 10%, for example, in the already quoted Japan.

Even more important is the fact that the rise in energy yield is on average 36%. This result is related to the fact that the CF of FPV ranges between 900 and 1200 hours with an average value of 1060, whereas the HPP capacity factor is on average 2950 hours.

Evaporation Reduction

A by-product of the FPV plant consists in a strong reduction of the evaporation from the water surface covered by the rafts which support the PV modules. This reduction factor depends on the raft design, and in our model (see Ref. [13]) we find that this reduction is approximately 80%.

As an example, we take the parameters of Longyangxia hydro-PV power plant. Let us imagine that the full PV plant was an FPV plant occupying a surface of approximately 10 km^2 and that we saved 2 m^3 of water per m^2, thanks to the reduction of the evaporation rate. The amount of water saved in 1 year would be 20 million m^3 which with an average head of 100 m gives rise to a production of 5.4 GWh, less than 0.1% of the full hydroelectric energy production.

This reduction is very relevant if we are in arid zones or, for example, for wastewater basins where an adequate water coverage can save from 10,000 to 20,000 m^3 per ha. However, in the case of HPP, saving water is not the main purpose, and we limited ourselves to some comments about the improvement of the HPP energy production due to the main amount of water in the covered area basins.

5. THREE LARGE FPV PLANT PROPOSALS

As shown in Section 4, the geographic potential is very large so that the possibility to install large FPV is overwhelming. Among the many possibilities, we analyze these three examples that can play the role of test cases.

Brazil: Balbina Dam

In Brazil, pilot projects of 5 MWp FPV plants in the lakes of the HPPs of Sobradinho (State of Bahia, Northeast

TABLE 8.6
Worldwide Analysis of Hydroelectric Basins.

	HPP			FPV			
	S_H, km^2	P_H, GW	E_H, TWh	S_{FPV}, km^2	E_{FPV}, TWh	$S_{FP}V/S_{Bv}$%	E_{FPV}/E_H, %
Africa	24,197	34.4	114.1	287	41.28	1.2%	36.2%
America Central	2899	7.7	23.0	64	9.24	2.2%	40.2%
America South	53,863	161	589.8	1342	193.2	2.5%	32.8%
Asia South East	22,929	44.8	143.4	373	53.76	1.6%	37.5%
Asia South—India	1081	17.6	65.5	147	21.12	13.6%	32.2%
Australia	1216	14	43.9	117	16.8	9.6%	38.3%
Canada	95,224	81	388.2	675	97.2	0.7%	25.0%
China	7454	333	1162.8	2775	399.6	37.2%	34.4%
Europe (North)	24,724	68	196.7	567	81.6	2.3%	41.5%
Europe (South)	2066	85.9	175.3	716	103.08	34.6%	58.8%
India	13,361	47.6	128.8	397	57.12	3.0%	44.3%
Japan	130	12	91.3	100	14.4	76.9%	15.8%
Middle East	10,775	15.87	19.8	132	19.044	1.2%	96.2%
Russia	84,761	51	186.6	425	61.2	0.5%	32.8%
Turkestan	14,582	12.7	51.7	106	15.24	0.7%	29.5%
USA	21,686	103	261.8	858	123.6	4.0%	47.2%
Total	**380,948**	**1089.57**	**3642.7**	**9079.8**	**1307.5**	**2.4%**	**35.9%**

FPV, floating photovoltaic; HPP, hydroelectric power plant.

region of Brazil) and Balbina (State of Amazonas, North region) were announced. The dam was established to provide a renewable electricity supply to the city of Manaus (250 MW power) but was considered by locals a controversial project from the start, due to the loss of forest and displacement of tribal home grounds.

The possibility to cover less than 0.1% of the basin doubling the installed power is shown in Fig. 8.6.

The HPP of Balbina produces 970 GWh yearly, and this should be compared with the energy of the PV plant of 364 GWh. This represents an increase in the energy production of more than 37% without any other environmental impact.

Ghana: Akosombo Dam

This is a particularly interesting situation since the hydroelectric power station of 1028 MW is managed by the Italian company ENI and also because the basin generated by this dam (2500 km^2) has problems of maintenance. The annual energy yield for a solar PV plant is very high (more than 1700 kWh per kWp) and the price of the produced MWh is below $40. In Figs. 8.7 and 8.8, two solutions are given for a 50-MWp FPV.

The choice between fixed floating plant or floating plant with tracking system is a matter of optimization. A fixed plant is simpler, and maintenance is easier; however, a tracking system allows a higher energy yield.

Here in Fig. 8.9 are two more extended fixed floating plants, of 100 and 200 MWp, respectively. At this size, further scale economies are possible. We remark that these plants cover less than 1% of the full basin generated by the Akosombo Dam.

The Akosombo Dam has generated a not negligible environmental impact: drought and algal bloom. In particular, we quote from Wikipedia [...] *The growth of commercially intensive agriculture has produced a rise in fertilizer run-off into the river. This, along with run-off from nearby cattle stocks and sewage pollution, has caused* eutrophication *of the river waters.* https://en.wikipedia.org/wiki/Akosombo_Dam [5]. *The nutrient enrichment, in combination with the low water movement, has allowed for the invasion of aquatic weeds (Ceratophyllum). These weeds have become a formidable challenge to water navigation and transportation.*

This problem is enhanced by the periodic droughts that have become a real challenge. However, several

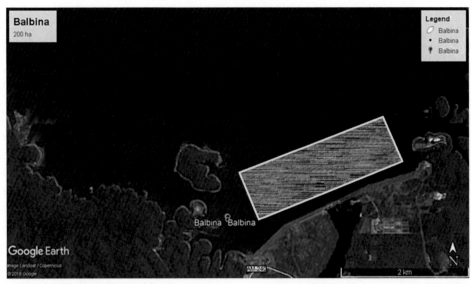

FIG. 8.6 Brazil: 280-MWp fixed floating photovoltaic plant near the Balbina Dam on a surface of 200 ha.

raft blocks= 418   area= 50 ha   power= 52.25 MWp

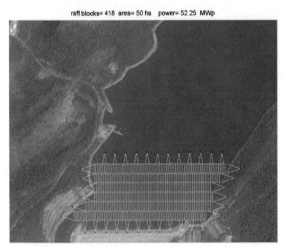

FIG. 8.7 Schematic view of a 52-MWp fixed plant.

bloks= 2564   panel= 123072   PV=320 W   Tot Power= 49.2 MWp

FIG. 8.8 50-MWp floating photovoltaic with tracking.

benefits can arise by the presence of a floating platform:

a) *Reduction of evaporation rate* in arid climate where evaporation in water reservoir wastes millions of cubic meters of drinkable water. A floating platform reduces the evaporation up to 90%, in the part of the lake covered by the PV plant.

b) *Algae bloom control.* Algae bloom is a planetary pollution problem. Floating plants reduce the algae bloom pollution by reducing the solar radiation and the algae proliferation mainly due to the combination of eutrophic water and photosynthesis.

c) *Water reservoir quality control.* This can be done with active systems like a flux of low-pressure laminar air as well as UT system (ultrasound transducers) with an improvement of the fish environmental habitat.

These advantages can contribute to a correct management of the large Akosombo basin.

Mozambique: Cahora Dam

Cahora Bassa hydroelectric plant in the north of Mozambique, with a basin of 2600 km^2, generates 2075 MW of electric power which gives energy to Mozambique, Zambia, and Zimbabwe.

In Fig. 8.10, the extreme east part of the basin (1 km^2 only) is shown, and on this surface it would be possible

FIG. 8.9 Two fixed floating PV plants of 100 and 200 MWp in locations near the Akosombo Dam.

FIG. 8.10 The branch of the Cahora basin which can host a 120-MWp FPV plant near dam.

to install a 1520-MWp FPV at very low cost, using the existing infrastructures. The energy production would reach 250 GWh/year at a cost below 40 $ per MWh. In this case, the limit to the FPV power is given by the orographic structure and by the possibility of having a grid connection in the vicinity.

6. CONCLUSIONS

Coupling of FPV and HPP allows a considerable increase in the RES energy production.

The cost of the FPV is reduced, thanks to the presence of infrastructures and the existence of a grid connection. The advantage related to the presence of a natural storage system is relevant.

Problems related to the managing of the plant, to the basin level variation, and to ice and snowfall have been analyzed and can be solved by simple and safe methods.

The cost of the FPV is comparable with that of a land-based PV plant and is further reduced, thanks to the presence of infrastructures and the existence of a grid connection. The advantage related to the presence of a natural storage system is quite relevant.

The strategy we suggest consists in installing an FPV power plant equal to the existing HPP power plant and in reducing the hydro turbines' energy productions during the sunny hours, and therefore maintaining the energy injected in the grid approximately constant.

The worldwide basin surface covered by FPV is only 2.4% but the increase in energy production is 35.9% raising the CF value from 3343 to 4450 hours.

This analysis can be extended to other situations and to smaller HPP basins where the CF factor is lower, i.e., around 2000. In this case, the benefits of the hybrid FPV-HPP coupling are more important, and the increase in energy can reach 50%.

REFERENCES

[1] BP Statistical Review of World Energy, BP Energy Outlook, London, 2018.

[2] Renewables 2018 Global Status Report, REN21 Steering Committee, 2018.

[3] International Hydropower Association, Hydropower Status Report, London, 2018.

[4] IRENA, Renewable Capacity Statistics 2018, International Renewable Energy Agency, Abu Dhabi, 2018.

[5] V. de Souza Dias, M.P. da Luz, G. Medero, D.T. Ferreira Nascimento, An overview of hydropower reservoirs in Brazil:current situation, future perspectives and impacts of climate change, Water 10 (2018) 592−610.

[6] R. Elrab, A. Ellah, Thermal stratification in lake Nasser, Egypt using field measurements, World Applied Sciences Journal 6 (4) (2009) 546−549.

[7] T. Nordmann, T. Vontobel, L. Clavadetscher, T. Boström, H. Remlo, Large Scale Hybrid PV hydroelectricity Production in floating devices on water, in: 24th European Photovoltaic Solar Energy Conference, Hamburg, 2009.

[8] Z. Glasnovic, J. Margeta, The features of sustainable solar hydroelectric power plant, Renewable Energy 34 (2009) 1742−1751.

[9] A. Sahu, N. Yadav, K. Sudhakar, Floating photovoltaic power plant: a review, Renewable and Sustainable Energy Reviews 66 (2016) 815.

[10] N. Martín-Chivelet, Photovoltaic potential and land-use estimation methodology, Energy 94 (2016) 233−242.

[11] S. Rehman, L. Al-Hadhrami, M. Alam, Pumped hydro energy storage system: a technological review, Renewable and Sustainable Energy Reviews 44 (2015) 586−598.

[12] G. Tina, R. Cazzaniga, M. Rosa-Clot, P. Rosa-Clot, Geographic and technical floating photovoltaic potential, Thermal Science 22 (Suppl. 3) (2018) S831−S841.

[13] R. Cazzaniga, M. Cicu, M. Rosa-Clot, P. Rosa-Clot, G. Tina, C. Ventura, Floating Photovoltaic plants: performance analysis and design solutions, Renewable and Sustainable Energy Reviews 81 (2018) 1730−1741.

[14] B. Ming, P. Liu, L. Cheng, Y. Zhou, X. Wang, "Optimal daily generation scheduling of large hydro−photovoltaic hybrid, Energy Conversion and Management 171 (2018) 528−540.

[15] R. Cazzaniga, M. Rosa-Clot, P. Rosa-Clot, G.M. Tina, Floating tracking cooling concentrating (FTCC) systems, in: 38th IEEE Photovoltaic Specialists Conference, PVSC, Austin, USA, 2012.

[16] M. Perez, R. Perez, C. Ferguson, J. Schlemmer, Deploying effectively dispatchable floating PV on reservoirs: comparing floating PV to other renewable technologies, Solar Energy 174 (2018) 837−847.

[17] C.I.A., The world facebook [Online]. Available: https://www.cia.gov/library/publications/the-world-factbook/.

[18] AQUASTAT, Dams: Geo-Referenced dDatabase, Food and Agricultural Organization of United Nations [Online]. Available: http://www.fao.org/nr/water/aquastat/dams/index.stm.

[19] Digital Water Atlas [Online]. Available: http://www.gwsp.org/products/digital-water-atlas.html.

CHAPTER 9

FPV and Environmental Compatibility

PAOLO ROSA-CLOT

1. INTRODUCTION

The managing of water resources is a planetary challenge, which must be dealt with by balancing the exploitation of renewable energy with the equilibrium of the environment. At present, many freshwater resources are polluted either because of their own formation (hydroelectric basins, quarry basins, and mining basins), because of the lack of adequate managing, or the problem of algae bloom, see Ref. [1]. The latter is mainly due to complex and general modifications such as climate changes and an excess of nutrients in water (see Refs. [2−4]). These problems are not limited to the freshwater resources but affect lagoons, salty water basins, and even the open sea where uncontrolled organic and inorganic waste disposal can generate critical situations [5]. Fig. 9.1 shows the consequence of Balbina Dam on the ecosystem of North Brazil [6].

This chapter addresses the problem of how a floating photovoltaic (FPV) plant interacts with existing water surfaces and analyzes the environmental impact and the benefits that can be derived by FPV of large power integrated with existing basins, natural or built as a consequence of industrial or agricultural activities. Particular attention will be devoted to:

1. Radiation, Evaporation and basin thermal equilibrium
2. Albedo and greenhouse effects
3. Integration with hydroelectric basins and relative environmental advantages
4. Quarry and mine basins
5. Wastewater basin management
6. Decommissioning of oil/gas platforms
7. Material compatibility problem
8. The fish and zootechnic equilibrium

2. RADIATION, EVAPORATION, AND BASIN THERMAL EQUILIBRIUM

The basin thermal equilibrium is determined by a complex balance between solar radiation, evaporation, and convective motions.

Actually, the first layer of 50 cm of the water basins absorbs about 50% of solar radiations so that the surface is warmer than deep layers; the infrared part of solar radiation is absorbed in the first few centimeters of water and the rate of the transmitted total solar energy at a depth z, $E_{tr}(z)$, is given by:

$$E_{tr}(z) = 0.44314 - 0.09213 \cdot \ln(z) \qquad (9.1)$$

where the depth z is given in m (relation valid in the range 0−10 m [7]).

This absorption is the energy source that determines the evaporation which balances the radiation input. This phenomenon is, however, strongly related to the wind and to the humidity rate so that the balance is far from being simple.

The evaporation rate is, after solar radiation, the parameter which mainly determines the thermal equilibrium of a water basin together with the solar radiation input, and we want to study how the PV floating system modifies it.

Let us assume that the basin depth is 5 m and solar radiation is 8 kWh/m^2 (a sunny hot day). If all the energy is converted into water's internal energy, the average temperature should increase by 1.37°C in 1 day. In practice, most of this energy is lost, mainly due to the evaporation and, in small part, to the radiation mechanism, so that the average variation of the temperature does not exceed the few tenths of °C during the sunny hours even if, in the absence of convective motions and of wind on the basin, the thin surface layer can increase its temperature by several °C.

The global energy balance and the evaporation rate are quite complicated to analyze, and several different approaches have been studied for describing this problem. In Ref. [1], the evaporation mechanism is discussed starting with the report [8,9] of the Australian Government Bureau of Meteorology, but several other models are available and many of them fit quite reasonably empirical findings [10−13].

Floating PV Plants. https://doi.org/10.1016/B978-0-12-817061-8.00009-9

FIG. 9.1 Balbina Dam in Brazil generates 2360 km² shallow water basin.

All these models use different input variables, but in general the most important are the following:

1. Relative humidity x ranging from 0 to 1.
2. Wind speed: u in m/s usually derived by u_{10} wind speed at 10 m above the water surface: u_{10} ranging from 0 to 20 m/s.
3. Water surface temperature: T_w in °C, in the range 5−25°C.
4. Net radiation at the water surface: Q in (kWh/d/m²) in the range of 3−8 kWh. Net radiation is the solar radiation minus the albedo effect plus the balance between incoming and outgoing longwave radiation. This depends on the cloudiness, but due to the complexity of a detailed calculation and to the limited effect of these corrections we will assume that Q is the incoming solar radiation.
5. Stored energy N (kWh/d/m²) with $\gamma = N/Q$ ranging from 0.5 to 1. This stored energy is mainly absorbed in the surface layers with consequences on the basin temperature profile.

Since there are many different models, they should be compared with and fitted to experimental findings. In Ref. [14], this work has been successfully performed and a check with experimental data of 50 days measurement has been done, with results which suggest the use of the more precise Penman model [12] depending explicitly on the five parameters quoted above.

Since the Penman model is rather complex, we have tried to use a simplified approach which starts from the remark that there are two terms, one related to the % of stored energy $(1 - \gamma)$ and the other on the relative humidity $(1 - x)$. Factorizing this dependence, the best fit to the Penman model with the five independent variables gives rise to a simple formula:

$$E_{vap} = (1 - \gamma)(0.353 + 0.0054 \cdot T_w) \cdot Q \\ + (1 - x)(0.20 + 0.08 \cdot u) \cdot (3.11 + 0.132 \cdot T_w) \quad (9.2)$$

which works quite well over a large range of parameters with errors that are less than 1%, much less than the model reliability. See Fig. 9.2.

The plot gives the evaporation rate in mm/day; evaluation is done for typical values of other parameters, in particular a sunny day has been chosen (radiation 6 kWh/m²) with a good transfer of energy to the water basin (heat storage 70%, i.e., 4.2 kWh/d/m²). Variation of these parameters does not affect results in a noticeable way [12]. On the contrary, variation of the vapor pressure and wind speed can change the evaporation ratio of a factor 10.

When water surface is occupied by a floating PV platform (see Fig. 9.3), we find that:

1. Wind speed at water surface below the raft is strongly reduced.
2. Thermal energy arriving at the water surface is reduced by a large factor due to the reflection and conversion efficiency of PV modules. Furthermore, infrared radiation due to the warming up of the PV

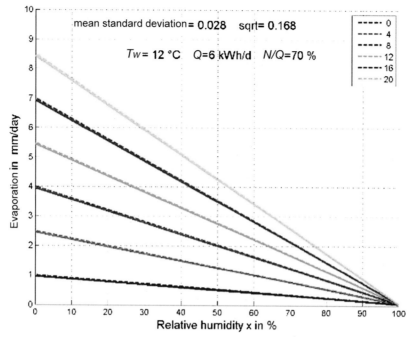

FIG. 9.2 Parameterization of the Penman model for different wind velocities (from 0 to 20 m/s). Dashed lines are our fit to the model.

modules is mainly emitted by the black PV surface rather than by the white rear part.

3. In the cavity, created between water and raft, the vapor pressure approaches the saturated vapor.

It is easy to see that the evaporation rate is reduced to a fraction of millimeter per day and so the evaporation reduction can be large. See also [15] where a reduction of 70% is suggested.

This is in contrast with a test done in Singapore (see Ref. [16]), where a reduction of the evaporation rate of only 30% is found; this result, however, has been obtained with a configuration rather different from the typical conditions of a floating raft. See also [17,18]. Evaporation mitigation of 65% was suggested, using empty drinking water bottles covering a pond in Ref. [19].

Our results agree with the detailed reference study [17,20] where an analysis for Lake Mead in Nevada has been done on the effect of an almost full coverage of the surface, using black or white large floating disks: Figs. 9.4 and 9.5. As evident, the coverage with circular floating disks (covering the full surface up to 80%) increases the surface water temperature, and the evaporation rate is strongly reduced to about 0.5 mm/day.

In order to give some general ideas on the reduction of evaporation rate, we have collected in Table 9.1 the main results for a basin with fixed water

FIG. 9.3 Raft structures which affect the evaporation rate in a different way.

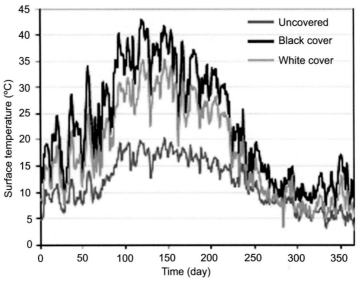

FIG. 9.4 Surface temperature [17].

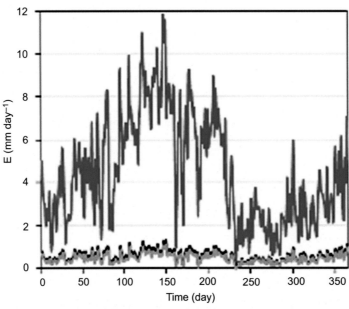

FIG. 9.5 Evaporation rate [17].

temperature at 20°C. The parameters chosen are not based on measurement, but on common sense hypothesis and should be improved by a test campaign. The equation given above has been used to estimate the evaporation and, as a consequence, the water saving and the reduction in percentage of the evaporation rate.

In the last column, the water saving with respect to an open surface, is given in m³/ha. As is evident, the presence of a floating structure gives an important

TABLE 9.1
Evaporation Rate for Different Configurations.

	Coverage %	Relative Humidity, x %	Wind u, m/s	Radiation/d, kWh/m²/day	% Storage, y %	Evaporation, m/y	Reduction, %	Gain, m³/ha
Open surface	0	0.3	10	8	0.8	1.74	0	0
Black disk	80%	1	0	7	0.8	0.54	69%	13,100
White disk	80%	1	0	5	0.8	0.48	72%	13,500
HDPE rafts	95%	0.8	1	2	0.5	0.36	79%	14,000
Korea	80%	0.7	3	4	0.5	0.84	52%	10,700
Singapore	90%	0.7	2	2	0.5	0.53	70%	12,600
Gable	95%	0.7	1	2	0.5	0.41	76%	13,400

benefit in terms of reduction of water losses due to the evaporation effect.

The problem about these data and the scatter of the final results are related to the experimental setup. If we work with PV modules in gable configuration, the wind reduction and the increase of the relative humidity are maximized, but these values can change dramatically with a different tilt of the modules or with a different shed and, actually, the configuration studied in Ref. [16] does not match the real structure of our floating raft.

The situation changes further if a cooling mechanism with water layer is operational on the PV panels. In this case, parameters are essentially the same used for an open water surface. When the cooling is switched on, in the sunny hours (typically 6—8 hours in a sunny day), a different analysis is necessary.

In that case, the water surface temperature is higher than the basin temperature, and this slightly favors evaporation. One should stress that cooling takes place mainly for conduction, leaving the evaporation mechanism untouched. However, a thorough experimentation is not available and a a more in-depth analysis is necessary; nevertheless, we might expect that in the absence of cooling, and using a compact raft system with a high-density modules packaging with low tilt angle, the evaporation rate is hindered by 70%—80%, whereas, if cooling is active, evaporation ratio is reduced by 60%—70%.

In deep basins the situation is more complex. In Ref. [21], Nasser basin is analysed and it is shown

that, although deep layers are not affected by seasonal variation, the behavior over the thermocline depends strongly on the solar radiation and requires a complex modeling.

See also the systematic work done on the Cumbrian lakes in Ref. [22]. If a partial coverage of the pond is active the energy balance changes deeply.

The energy entering the pond is reduced by about 90% (we always refer to the part covered by the PV floating plant) and, even if evaporation is strongly reduced, the balance is negative, and the pond temperature is slightly decreased. If the pond is fully covered, some changes in the temperature are possible, but the conduction mechanism becomes important and helps to maintain stable temperature levels.

The reduction of solar radiation arriving on the water surface can strongly affect the thermodynamic equilibrium.

Actually, at variance with white/black disks of reference [15], the coverage of a large basin surface with a gable structure forbids the radiation from reaching the water. In this case, 5% of the radiation is reflected, 15% is converted into electricity, and 80% goes to warm the PV modules which reach a thermal equilibrium by radiating in the far infrared region mainly from the top part. The bottom part has a lower temperature (5°C less) and a lower emissivity being usually of white plastic.

So the thermal input is strongly reduced and the temperature of the basin should lower considerably. However, the water evaporation rate is strongly reduced

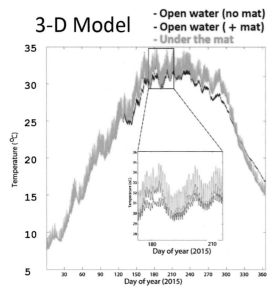

FIG. 9.6 Surface temperature with and without mat [23].

as well, and a balance should be done by comparing the two effects.

Using our formula of Eq. 9.2 we find that the evaporation rate is reduced by a factor 10, and this can qualitatively match the reduction of thermal input, thus leaving the situation unaltered.

Several simulations have been done for this topic. In Fig. 9.6 results of a model of basins with mat covering only 10% of the surface, is shown together with the simulation of hourly temperature near the surface with and without mat [23].

The conclusion is that the temperature in the layer immediately below the mat is slightly increased of approximately 1−2°C. Other papers quote analogous results, but research in this sector is in rapid evolution both with simulation and experimental tests. See also Refs. [17,24].

It is not simple to perform a detailed analysis of the thermodynamic balance. A mix of warm water, coming from the cooling, with the water of the basin surface can slightly modify the water surface temperature, but we have not enough data for doing a reliable model.

Experiments are necessary using large covered surfaces in order to get quantitative results. Work is in progress. In conclusion we can state that:

1. Radiation input on water basin is substantially reduced because of the floating structure; however, also evaporation rate is reduced so that the net effect on the temperature is limited and depends on specific local conditions.

2. Evaporation rate reduction favors water saving, which in equatorial regions can reach 15,000 m^3/ year for hectare.
3. The presence of FPV does not modify in a sizable way the basin thermodynamic equilibrium.
4. If cooling is active the gain in efficiency is sizable, but there is also an average increase of the basin temperature.
5. The lack of systematic experiments on large covered surfaces does not allow a definitive conclusion of this last issue.

3. ALBEDO FOR PV PLANTS AND GREENHOUSE EFFECT

It is interesting to analyze the albedo effects of PV plants using the results of the analysis presented in Chapter 3.

PV modules are built in order to absorb the solar radiation almost completely, and surface glass is often laminated in order to minimize reflection and possible glare. In general, we can assume that the absorption of sunlight (diffuse and direct) ranges between 95% and 97%.

This has a consequence on the global energy balance. The solar radiation is partially converted into electricity (which after use becomes heat at low temperature). This portion is typically 15%−20% of the incoming radiation. The remaining part (85%−80%) becomes heat, and the PV modules warm up and reach a thermodynamic equilibrium when the emitted infrared radiation balances the solar radiation input.

This simple energy balance may usefully be described with a formula. If the land albedo is α_{Land} and the albedo of PV module is α_{PV}, and the electric energy produced is E_{electric} due to a PV efficiency of η, the increase of energy contributing to the greenhouse effect, ΔE_{GH}, is:

$$\Delta E_{\text{Gh}} = \frac{E_{\text{electric}}}{\eta} \left(\alpha_{\text{land}} - \alpha_{\text{PV}} \right)$$

Looking at the installed PV capacity at present worldwide of 400 GW with an annual energy yield of 445 TWh/year, and assuming an average land albedo of 0.3, PV albedo of 0.05, and an average efficiency of 14%, this effect will increase the energy balance by about 800 TWh.

This value is very high, but should be compared with the global earth energy input which is 174,000 TW, that is 1.5 10^9 TWh/y. So, the yearly contribution is only 0.5 ppm, which is small, but not negligible, especially if considering that it has a cumulative effect. This contribution should shift the final thermal balance and is going to augment because of the great investments in the PV solar sector.

Microclimate changes have been observed, due to the installation of large PV systems and sometimes the local effects can be positive (see, for example, [18]), but, in any case, the global change induced on the earth equilibrium results in an increase of the greenhouse effect.

In contrast, nothing happens if an FPV plant is used. In this case, the low water albedo is approximately equal to the PV modules albedo so that the global balance remains unchanged.

In conclusion, the albedo effect on land-based PV plants alters the energy balance giving net contribution to the warm-up of the planet; this effect is zero for FPV plants.

FIG. 9.7 Drought due to microclimate changes induced by dam construction (China) [27].

4. MITIGATING THE IMPACT OF HYDROELECTRIC POWER PLANTS

The possibility to organize basins with grid connection and, sometimes, with a storage pumping system, has been extensively analyzed in Chapter 8, and the advantages in energy production are quite evident. Furthermore, FPV plants mitigate the environmental problems opened by the hydroelectric power plant (HPP) [25]. These problems are numerous, and in the following we compare three different synthesis for Assuan Dam, China HPP, and Brazil Andean dams.

Assuan Dam on the Nile River. The impact of Assuan Dam has been studied and a list of the main problems given in Ref. [26] is given below:

1. excessive sedimentation in the water upstream of the dam and erosion of those downstream;
2. increase of the salinity of the delta (with the decrease of the strength of the Nile, the salty waters of the Mediterranean have advanced along the river);
3. decrease in fishing productivity along the river and disappearance of species that migrated along the Nile;
4. migration of marine animals into the river due to increased salinity, increase in groundwater levels in fields close to the river with consequent water stagnation (which in turn causes the spread of fungal pathogens);
5. decrease in the fertility of the land downstream of the dam because without flood the silt does not reach the ground.

China HPP managing. More recently, China water authorities have published critical comments on HPP managing [27]. According to the government, the concentration of rain during summer and autumn will increase, overwhelming rivers in the south, while long dry winters will increase. The impact of climate change

in combination with the boom of dam-building in China, will likely result in the following:

1. Dam mismanagement exacerbates drought conditions. In 2011, central China was hit by its worst drought in five decades, which sharply reduced the power generation of the Three Gorges Dam giving rise to tension between upstream and downstream users, especially during the dry season (Fig. 9.7).
2. Reduced runoff decreases water availability for dams and hydropower generation. According to the government's climate change assessment, since 1950s, more than 82% of glaciers have been in a state of retreat; and the pace has accelerated since the 1990s.
3. Sediment buildup behind dams reduces deltas' capacity to mitigate sea level rise. Sea level rise combined with sediment buildup behind dams poses a threat to China's three major deltas, the Yellow, the Yangtze, and the Pearl.

Brazil and HPP Impact

In Brazil, renewable energies account for more than 35% of electric energy production, thanks to the large HPP dams, and the forecasts carried out by the government are very promising. In fact, this percentage should increase to 47% in 2040 [28].

Actually, the HPP sector has been in expansion for several years, and HPPs are distributed on all the territory so the problem of a correct managing of the environmental impact is dramatic (see, for example, Fig. 9.1 for effect of Balbina Dam on the territory). This situation is not unique, and the construction of six Andean dams will have extensive impacts on Amazon fluvial ecosystems that need to be addressed in regional infrastructural development plans. Fig. 9.8 summarizes the principal impacts considered here,

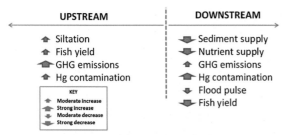

UPSTREAM	DOWNSTREAM
⬆ Siltation	⬇ Sediment supply
⬆ Fish yield	⬇ Nutrient supply
⬅ GHG emissions	⬆ GHG emissions
⬆ Hg contamination	⬆ Hg contamination
	⬇ Flood pulse
	⬇ Fish yield

KEY
⬆ Moderate increase
⬅ Strong increase
⬇ Moderate decrease
⬇ Strong decrease

FIG. 9.8 Expected environmental impacts above and below Andean dams [29].

indicating their relative magnitude and importance to ecosystem structure and functions.

All these problems in three very different contexts are quite similar and their negative impact can be attenuated by the integration with FPV plant covering a small percentage of the hydroelectric basin surface.

• *The floating structure reduces the evaporation of the free surface of the water, keeping the volume of stored water.* In average, this gives a water saving of $1-2 \text{ m}^3$ for each m^2 of FPV surface.

• *The floating PV array reduces algae growth.* So it can contrast negative algae bloom and methane production due to the decomposition of organic matter.

• *The floating structure reduces the formation of waves and, thus, the erosion of the banks of the reservoir.* This is limited to the covered surface which normally is a small percentage of the basin surface, but is particularly important since the FPV plant should be normally located not far from the dam structure, where these phenomena are more dangerous.

• *A floating PV system installed in the reservoir of a hydropower plant does not require investment in transmission infrastructure, since the existing grid can be shared.* This depends of course on the local structure

and grid connection, but limiting the power of the FPV plant to the maximum hydroelectric power, the energy production can be optimized.

• *A floating FPV can limit the exploitation of the water resources.* Since the FPV produces energy for 1,200−1,800 hours/year, we can manage the hydroelectric power during the FPV production time, reducing the water flow and optimizing the basin water resources.

Let us insist on the Balbina reservoir, which has generated many problems for local inhabitants and the environment of the Amazon rainforest [29]. Its surface area, a large fractal in the Amazon forest, is 2,360 km^2 with an average depth of 7 m; the HPP has a power of 250 MW. Problems of fouling, methane emission due to the organic product decomposition are dramatic.

An FPV of 250 MW would cover only 2.0 km^2, i.e., less than 0.1% of the available surface; in addition to the generation of electricity, the plant can be used as a platform and can power equipment for managing the water quality and the fouling issues See Fig. 9.9.

5. QUARRY AND MINES BASINS

In many European regions, there are sand quarries that, after exploitation, are in general abandoned without any suitable managing. In Fig. 9.10, the map of an industrial zone, 6 km from the center of Brescia (Italy), is shown. All the basins are sandpits, exploited and exhausted, and have a surface of approximately 1.5 km^2. Their total coverage should allow the construction of 200 MWp FPV with a yearly production of electric energy exceeding 250 GWh.

This situation is characteristic of the whole Po valley. In Figs. 9.11 and 9.12, two large basins are shown: the first one in the center of Milan, the second 20 km

FIG. 9.9 Balbina fractal structure. On the left a zoom of the dam zone with a 250 MWp FPV.

FIG. 9.10 Quarry basins near Brescia. The available water surface is more than 1.5 km².

West. The yellow perimeters include areas of 10 ha and 60 ha, respectively, but many other basins are near or anyway in the zone.

The same situation can be found in many locations in Italy, and their correct management is very important, since, most of the time, they become abusive waste disposals or, in any case, they are dangerous areas.

A similar situation happens in many part of central Europe. Fig. 9.13 shows a sand quarry managed by the company SMALS in Holland. SMALS is the leader of the sand market in central Europe and manages many square kilometers of sandpits.

In a complete different context, several research groups have begun to analyze the situation of abandoned mines. See Refs. [30–32].

This research group works mainly on the problem of abandoned mines, see Figs. 9.14 and 9.15 where exhausted diamond and gold mines are shown. The solution of covering part of the waste basins can be useful for the control of environmental conditions and for producing energy necessary for the management of abandoned mines.

In all these situations, land degradation due to the excavation activities can be limited by installing FPV plants. The low cost of the FPV allows the production of energy at a very low price and guarantees the income necessary to manage these very large water surfaces.

6. OIL PLATFORM DECOMMISSIONING

We are rather skeptic with offshore solutions. Open oceans are a challenging environment, and large waves and windstorms are able to destroy any floating structure except very expensive and well built ships or platforms.

However, recently, an environmental problem has arisen due to the large number of offshore platforms for extracting hydrocarbons. Oil gas platforms have a typical life cycle of 25 years so that the problem of their decommissioning is very important. There are more than 6,500 platforms worldwide, and their aging urges a decommissioning-recycling program. In the Adriatic Sea

FIG. 9.11 Cave Park (Milan), 10 ha.

FIG. 9.12 Basin at Milan East, 60 ha.

only, 127 platforms exist, mainly managed by ENI, and most of them are at the end of life; furthermore, more than 600 exist in the North Sea and North East Atlantic and almost all are at the end of their life cycle [33].

The main idea explored in Ref. [33] is to use the existing gas/oil platform as support for a hydrogen or methane synthetic production powered by the energy produced with floating plants and to use existing pipes and connection to the mainland for sending the produced gas to the end user.

Therefore, it is possible to produce green hydrogen through electrolysis, with equipment located on the platform [34].

Alternatively, a reverse osmosis plant can be located on the platform in order to produce freshwater from seawater.

The natural solution of using FPV plants faces two problems: the possibility of strong wind gust (the Adriatic is a rather quiet region, but sometimes Bora wind can overcome the 100 knots speed) and the presence of high waves. High waves are a seldom phenomenon in the Adriatic, but, even there, 6-m high waves can occur: in a period of 24 years, three events of 6 m waves were registered and 18 with waves in the range of 5−6 m. In Northern Europe, wave heights of 10 m are not uncommon, and so this problem should be addressed from the very beginning.

About wind gusts, the gable solution proposed in Chapter 4 is, without doubt, the best one. Even if it slightly reduces the solar harvesting, its characteristics of robustness, reliability, and resistance to strong winds are what is needed in a marine environment.

The wave problem, however, cannot be easily solved for two reasons: compactness of a PV large platform exposes the structure to the impact of large masses of water and any effort to resist a strong storm with large waves is useless.

The only possibility is to sink the platform when waves overcome a moderate height, which should be chosen at 1 m (in this way 95% of the sea conditions satisfies this constraint).

The sinking system is based on the possibility of introducing water in the pipes, which are the base of the floating structure, and of sinking the platform under a water layer of several meters, from 5 to 10, depending on the wave height.

The basic solution is shown in Fig. 9.16. The buoyancy of the platform is about 1,500 kg per raft, but, as

FIG. 9.13 SMALS (Nijmegen) sandpit.

FIG. 9.14 Victor Mine in Canada (diamonds).

FIG. 9.15 Canada Gold mine.

the volume of each pipe is about 3 m³, it is sufficient to flood water in the pipes in order to gently submerge the full structure.

Mooring cables, if suitably positioned at large distance, allow for a vertical movement of a large platform of 10 m or more, without changing their position or changing length and strength.

When the storm abates and waves are below a fixed height (1−2 m), the platform can emerge by inflating air in the pipes and pumping out most of the water.

The dimensions of these FPV plants can be quite large. In Ref. [33], a first hypothesis is done for four platforms in the Adriatic Sea, chosen with low sea depth, in order to minimize wave impact and favor the coupling to fisheries and aquacultures: the suggestion is to install 58.5 MWp of FPV systems. Assuming that each oil platform is equipped with a 15 MWp FPV, the sea surface occupied by the plant would be approximately 10 ha for each platform and this should produce approximately 20 GWh of electric energy per year. This energy should be mainly used for local hydrogen production or for synthetic methane production (Fig. 9.17).

7. THE HARMFUL ALGAL BLOOMS

The term "harmful algal blooms" (HABs) covers events that are caused by microalgae and threaten to have a negative impact on human activities.

The excess of nutrients may originate from fertilizers that are applied to land for agricultural or recreational purposes as well as for inadequate wastewater treatment. When phosphates or organic wastes are introduced into water systems, this can cause an increase in the growth of algae favored, during the summer, by the intense solar radiation; but algae are short-lived, and their decomposition consumes the oxygen in the water, resulting in hypoxic conditions and in toxic products being dissolved in water.

This problem has been widely analyzed in Ref. [1] (Chapter 10) and we give here only a short synthesis summarized in the SWOT analysis for the most important techniques used.

	Strength	Weakness	Opportunity	Threats	1/10
Chemical precipitation	Very effective	Adds chemical products	Suitable for confined industrial basins	Long-term effects out of control	4
Air flotation	Effective on any dust or biofouling	Limited efficiency		High energy consumption	7
Continuous laminar flow inversion and oxygenation	Beneficial effects on the whole ecosystem		Oxygenates the water body from top to bottom as well as feeding aquatic weeds and algae	Gradual introduction of the system to avoid fish kills	9
Electro flocculation	Acts as coagulant and flocculant	Adds metals to the water Energy expensive	Favors the remediation in confined very polluted pond	Long-term effects out of control	3
Sonication	Strongly hinders algae development	Effects on larvae should be evaluated	Long range action with very low energy consumption		8

FIG. 9.16 Oil/Gas platform in the Adriatic Sea and floating photovoltaic element with sinking system.

FIG. 9.17 Oil platform with floating photovoltaic plant.

The two techniques which show a small environmental impact, as well as limited energy consumption, are air flotation-oxygenation and sonication.

Flotation-oxygenation is free from threats except the fact that, even if it is easy to prepare a system which produces microbubbles of defined size, the distribution on a large area implies long pipe systems and using part of the electric energy produced by the FPV plant.

The use of an aeration/oxygenation system should be introduced starting with short times of application with a gradual increase so as to avoid fish kills: this is due to the fact that oxygenation changes the ecosystem natural balance.

Low-frequency ultrasounds transponder (UT) can be used for local treatment in a limited space. In this case, UT breaks the cells membranes and favors the coagulation, whereas air flotation brings to the surface the resulting sludge and can be used coupled to mechanical action and a barge-skimmer system. A series of UT, working at three different frequencies in the range 20−100 khz, acts with intense density energy on a limited confined volume where basin water is pumped at very low speed. In these volumes, an air

FIG. 9.18 Dense algae bloom: when skimmer is necessary.

flocculation process takes place and allows a removal of dead algae.

In Fig. 9.18, three situations are shown where a shock therapy might be necessary. In this case, high-frequency ultrasounds might be less effective since they are quickly damped by the sludge, and the use of low frequency UT with air flocculation system might be needed.

8. MATERIALS COMPATIBILITY [35]

FPV plants occupy large surfaces of freshwater basins or of the sea and, even if the coverage is, in most cases, a small percent, the impact of floating structures on water quality has to be carefully evaluated considering the three basic materials used for building rafts:

- High-density polyethylene (HDPE)
- Galvanized steel
- Aluminum

Actually, the direct contact with water is mainly through HDPE pipes supporting structures in galvanized steel or through rafts fully built in HDPE. Galvanized iron (or aluminum) is not in direct contact with water, but for several reasons such as rain or waves these structures and the PV modules can be wetted by water and can release small quantities of materials dissolved in water.

The metal structure type is usually common and cheap steel (S235 JR) that must be galvanized in order to guarantee the life cycle of the structure. The durability is guaranteed by the vendor, and a guarantee of 25 years or more is not unusual for this class of material. The steel must be hot-dip galvanized at a temperature of almost 500°C.

The protection of the coating depends mainly on the thickness of the zinc layer. Year after year the zinc dissolves at a very slow rate, but this does not create any problem. The release caused by the structure is so slow that it only slightly changes the concentrations naturally present.

Aluminum is used for the PV modules frame and, sometimes, for modules support both in floating plants

and in PV plants on land. A typical wearing out of metals for 1 MWp on land is 10−12 tons of anodized aluminum for PV frames and 30 tons of galvanized iron for modules support.

On a floating plant, these numbers are quite different and for 1 MWp, the quantities involved can be roughly estimated as follows:

1. HDPE pipes: from 1300 to 5000 m of pipes (depending on the technical solutions adopted)
2. Galvanized steel: beams (omega) plus legs and flaps, 30−40 tons.
3. Anodized aluminum: module frames, 10 tons.
4. Other components of PV modules (glass, eve, etc.), 50−60 tons.

The structure in galvanized steel can be substituted by a structure in aluminum. This is much lighter than iron, but its tensile strength is much lower so that the final weight would be approximately the same.

High-Density Polyethylene

Water transport is today based on several possible solutions, but HDPE is the widespread material for low-pressure water conduits for residential use and has now completely substituted galvanized iron pipes. The benefits are many, in particular a life cycle of around 50 years and total compatibility with the environment, plus simplicity in handling.

The only limit is in its resistance to UV radiation which after a few years reduces its ability to withstand shocks and mechanical strain. This limit is contrasted by adding carbon black to the plastic, and in our project the HDPE pipes are black and completely covered by PV modules so that the problem is solved from the very beginning.

Finally, pipes suppliers offer a product certified for drinking water following European standards.

Zinc

Producers claim that the corrosion of a galvanized steel beam depends on the environment and on corrosive elements in the atmosphere. In standard conditions the coating protection can last more than 100 years.

However, in order to be conservative, we will assume that in 25 years the corrosion consumes 10% of the coating when in contact with freshwater.

Hot zinc coating gives protection also for plants working in seawater or exposed to salty water spray: the factors that influence the corrosion of zinc in freshwater also apply to seawater.

In theory, a high level of chloride (as is the case for seawater) suggests a high corrosion rate. However, other salts are present in seawater, and the presence of magnesium and calcium contrasts zinc corrosion. Furthermore, all metal structures are suspended at least 8 cm or more above water, depending on the size/diameter chosen for the floating HDPE pipes, so that no continuous direct contact exists between galvanized iron and water.

Let us now estimate the quantity of zinc which can be dissolved in water in 1 year for 1-MWp plant; the steel mass of a 1-MWp platform is approximately 100 tons. Assuming a 4% of zinc, we find a mass of zinc of 4000 kg. This mass undergoes a corrosion process of few microns per year (values of 3—4 microns are suggested as typical in external environment) so that the dispersed zinc mass for year is about 20 kg.

This value should be compared with the total water exchange in the basin in 1 year. There is a great variability in this parameter: in hydroelectric basins the water exchange is quite rapid as well as in wastewater treatment plants, whereas in water reservoir the water can be stored for several months. A further parameter is the percent of the basin surface occupied by the floating plant.

The zinc concentration should be below 50—60 µg/L as suggested in medical studies.

Any excesses (over 50—100 µg/L) as well as shortages (below 1 µg/L) would lead to negative effects.

Plot in Fig. 9.19 shows the zinc concentration level that is optimal for biological activity and determines the range where it must be kept.

This is a qualitative plot based on the concept of biological activity which can be defined as the capacity of a molecule to generate a biological result.

Values of 5—22 mg in 1000 L of water have been quoted for zinc concentration in different world areas. The zinc intake for a typical North American diet ranges between 10 and 15 mg/day, whereas for Northern Europe values ranging from 12 to 16 are quoted.

Several recent papers about the zinc presence in diets in developing countries suggest that attention should be given to increasing zinc in children diet, since this should increase their immunological defense.

In conclusion, it is estimated by WHO that a nutritional deficiency of zinc may be affecting nearly two

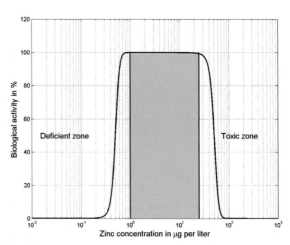

FIG. 9.19 Plot of biological activity in percentage versus zinc concentration in µg/liter.

billion subjects. Zinc therapy has reduced the mortality of children who developed diarrhea, and millions of lives are being saved worldwide.

Aluminum

Aluminum, at variance with zinc, is not essential for life. The concentration of aluminum in natural waters is below 0.1 mg/L; this low value is due to the very low solubility of this metal in fresh and salty waters.

Renewable energy systems (RESs) offer a large market opportunity for aluminum.

This is due to

- the large area of energy collection (e.g., module or collector surface) of RES
- the requirement of solar-directed installation (e.g., mounting frames of solar PV plants or of thermal solar panels)
- the expected dynamic market development (high expansion targets in many countries)

Water is sometimes treated adding aluminum salts in order to become drinkable, but aluminum levels do not exceed the concentration of 0.1 mg/L so that very small quantities of aluminum are absorbed by drinkable water.

Furthermore, it is not clear how an excess of aluminum is absorbed by the human body and for these reasons no limits have been imposed up to now in the use of aluminum cans for drinkable liquids. Therefore we can safely use aluminum for PV modules frames and for their supports.

In conclusion, the FPV plants built with HDPE pipes and galvanized steel are completely safe and have no effect on the quality of the water.

Steel	**Aluminum**	**HDPE**
• Traditional material suitable for analysis and engineering. • Rust and durability issue. • Needs repetitive maintenance/control. • Environmental pollution caused by rust. • 7.8 specific gravity	• Suitable for structural analysis. • Fair durability in wet and humid environment. • 2.8 specific gravity for workability. • Strength of welded parts not reliable.	• Fair durability in wet and humid environment. • Good durability in saline environment. • Lightweight material with 1 specific gravity. • Very low strength per unit of weight.

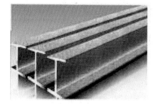  

FIG. 9.20 The three typical materials.

In Fig. 9.20, a synthesis of the three materials is given.

9. FPV AND FISH AND ZOOTECHNIC EQUILIBRIUM

The impact of FPV on external environment is very important since control and managing of water basins can be coupled to a set of positive biological effects.

When FPV is combined with aquaculture, we speak of aquavoltaics [36]. The goal of aquavoltaics is the efficient use of water for both food and energy generation. While solar panels, above the water or on its surface, provide the electrical energy, the aquatic organisms living within the water below provide a sustainable food source. The concept of aquavoltaics has both scalability for industrial-sized farms and the capability for off-grid remote location individual farms. This activity does not concern only fish, but also human activities such as fisheries or shrimps farms (see Ref. [37]).

Birds

Another impact is on birds which consider the floating platform as a useful support for nesting. As an example see the seagull nest on our floating platform in Suvereto. See Fig. 9.21.

Bird activities and specifically bird droppings can pose some problems, but in any case it is a proof that FPV does not affect bird fauna at all. Fig. 9.22 shows a photo of bird droppings, and many examples can be

FIG. 9.21 Seagull nest in Suvereto FPV.

FIG. 9.22 Bird droppings on a floating photovoltaic plant of the SERIS test bed.

FIG. 9.23 Cirata reservoir, 83 km².

FIG. 9.24 Detail of rafts on Cirata lake, 62 km².

found in analogous situations. This dramatically hinders the energy harvesting.

This raises the problem of controlling birds' activities without harming them, and a water veil generated by a cooling system is the best way for cleaning PV modules, but can also be used for dissuading birds from resting on the PV modules.

Fish

As to ichthyic fauna, a positive response to the installation of FPV has been observed. Our plant in Colignola was on a basin used for sport fishing, and carps populating the basin enjoyed staying under the platform to the disappointment of the fishermen. This was mainly due to the partial shielding from the sun and to the presence of small algae formation under the HDPE pipes. These positive effects have been studied in previous works and in different contests, see, for example, Ref. [38].

But the true question is if this natural cohabitation can be transformed in an advantageous investment. An example can be found in Indonesia where the Jakarta capital is supplied by many reservoirs. The most important are located in the south east and are the Jatiluhur and Cirata reservoirs which are artificial and equipped with large HPPs.

One of the most interesting characteristics of these basins is the important fishing activity and the presence of a lot of rafts floating on the Cirata lake used for fishing and housing.

See Figs. 9.23 and 9.24. In this case, coupling with FPV would be very inexpensive and fruitful.

It is quite evident that, with a suitable organization of these basins, the huge number of rafts used for fish farming could be integrated with FPV system without modifying the current activities, but adding to these the energy production.

10. CONCLUSIONS

The impact of FPV on environment is complex but globally positive since it can mitigate the negative impact of other invasive human activities.

We summarize the main results of our analysis:

- Since albedo effect is not altered by the presence of FPV plant, FPV does not give any contribution to the global warming. This is an advantage compared with land-based plants which normally reduce local albedo and increase the earth energy budget.
- FPV plants strongly reduce the evaporation on water reservoirs, and in this way protect water resources and save typically 15,000 m³/year for each ha of FPV plant. The basin temperature remains substantially unaltered.
- Coupling of FPV and HPP allows a considerable increase in the RES energy production. The cost of the FPV is reduced, thanks to the presence of infrastructures and the existence of a grid connection. The advantage related to the presence of a natural storage system is relevant.
- Even if some criticism has been raised about the real advantages of this coupling (see Ref. [39]), most of the authors conclude that the HPP and FPV together guarantee scale economies and a reduction of the environmental impact. See also Refs. [40,41].
- Quarries and mines. The advantages are overwhelming and an investment in this sector is important from the economic and environment aspects.
- Oil platform decommissioning. Notwithstanding the problem of large waves, the possibility to couple the huge investments necessary for a correct oil platform decommissioning to solar energy production suggests the use of FPV plants. These, however, should be submerged when waves are above 1 m, and this requires a further development of the FPV technology.

- Algal bloom. This distressing problem is on the increase, but can be partially managed by FPV. Of course, FPV does not eliminate the problem but helps in reducing its impact in simple and efficient way.
- Materials used are compatible with environment requirements and there is also the great advantage of no land use and of a very simple decommissioning.
- Fish and zootechnic equilibrium. There is evidence of a good integration with aquatic fauna and even birds seem not to be bothered by FPV structures. On the contrary, gentle modes for avoiding bird droppings and their impact are possible.

REFERENCES

[1] M. Rosa-Clot, G.M. Tina, Submerged and Floating Photovoltaic Systems, Modelling, Design, Case Studies., Elsevier, Academic Press, London, 2017.

[2] I. Sanseverino, D. Conduto, L. Pozzoli, S. Dobricic, T. Lettieri, Algal Bloom and its Economic Impact, 2016. JRC Technical report.

[3] L. Shen, H. Xu, X. Guo, Satellite remote sensing of harmful algal blooms (HABs) and a potential synthesized framework, Sensors 12 (2012) 7778–7803.

[4] R. Kudela, et al., Harmful Algal Blooms. A Scientific Summary for Policy Makers, IOC/UNESCO, Paris, 2015.

[5] M. Fossi, T. Romeo, M. Baini, C. Panti, L. Marsili, Modeling plastics and mediterranean whales, Fontier in Marine Science 4 (2017).

[6] International Hydropower Association, Hydropower Status Report, 2017. London.

[7] M. Rosa-Clot, P. Rosa-Clot, P. Scandura, G.M. Tina, Optical and thermal behaviour of submerged PV solar panel: SP2, Energy 32 (2011) 17–26.

[8] Australian Government Bureau of Meteorology, "Average evaporation Australia," [Online]. Available: http://www.bom.gov.au/. [Accessed 24 November 2016].

[9] S.A. Water, SA Water Wastewater Treatment Plants and Catchments, 2012. Adelaide.

[10] J. Monteith, Evaporation and surface temperature, Quarterly Journal of the Royal Meteorological Society 107 (1981) 1–27.

[11] D. McJannet, I. Webster, M. Stenson, B. Sherman, Estimating Open Water Evaporation for the Murray-Darling Basin, 2008.

[12] H. Penman, Natural evaporation from open water, bare soil and grass, Proceedings of the Royal Society of London Series A 193 (1948) 120–145.

[13] J. Valiantzas, Simplified versions for the Penman evaporation equation using routine weather data, Journal of Hydrology 331 (2006) 690–702.

[14] F. Bontempo, G. Tina, A. Gagliano, Study of Evaporation Reduction in Water Basins with Floating Photovoltaic Plants (to be published), 2019.

[15] M. Taboada, L. Cáceresa, T. Grabera, H. Galleguillosa, L. Cabezab, R. Rojasc, Solar water heating system and photovoltaic floating cover to reduce evaporation: experimental results and modeling, Renewable Energy 105 (2017) 601–615.

[16] G. Melvin, Experimental Study of the Effect of Slar Floating Panels on Reducing Evaporation in SIngapore Reservoir, National University of Singapore, Singapore, 2015.

[17] M. Aminzadeh, P. Lehmann, D. Or, Evaporation suppression and energy balance of water reservoirs covered with self assembling floating elements, Hydrology and Earth System Sciences 22 (2018) 4015–4032.

[18] O. Gaikwad, U. Deshpande, Evaporation control using floating PV system and canal roof top solar system, International Research Journal of Engineering and Technology 4 (4) (2017) 214–216.

[19] M. Hassan, W. Peirson, Evaporation mitigation by floating modular devices, Earth and Environmental Science 35 (2016).

[20] X. Yao, H. Zhang, C. Lemckert, A. Brook, P. Schouten, Evaporation Reduction by Suspended and Floating Covers: Overview, Modelling and Efficiency, University of Queensland, Griffith, 2010.

[21] R. Elrab, A. Ellah, Thermal stratification in lake Nasser, Egypt using field measurements, World Applied Sciences Journal 6 (4) (2009) 546–549.

[22] A. Folkard, Temperature structure and turbulent mixing processes in Cumbrian lakes, North West Geography 8 (2008) 42–50.

[23] G. Schladow, Reservoir modelling and quantifying evaporation reduction effect of floating solar, in: University of California, Davis, 2017.

[24] M. Magee, C. Wu, Response of water temperatures and stratification to changing climate in three lakes with different morphometry, Hydrology and Earth System Sciences 21 (2017) 6253–6274.

[25] B. Ming, P. Liu, L. Cheng, Y. Zhou, X. Wang, Optimal daily generation scheduling of large hydro–photovoltaic hybrid, Energy Conversion and Management 171 (2018) 528–540.

[26] A. Biswas, C. Tortajada, Impacts of the high Aswan dam, in: Impacts of Large Dams, Springer, 2012, pp. 379–396.

[27] A. Botelhoa, P. Ferreira, F. Lima, L. Costa Pinto, S. Sousa, Assessment of the environmental impacts associated with hydropower, Renewable and Sustainable Energy Reviews 70 (2017) 896–904.

[28] K. Yan, Climate change & hydro: mutually damming, China Water Risk (2012).

[29] V. de Souza Dias, m. da Luz, G. Medero, D. Ferreira Nascimento, An overview of hydropower reservoirs in Brazil: current situation, future perspectives and impacts of climate change, Water 10 (2018) 592–610.

[30] B. Forsberg, J. Melack, T. Dunne, R. Barthem, M. Goulding, R. Paiva, M. Sorribas, L. Silva, S. Weisser, The potential impact of new Andean dams on Amazon fluvial ecosystems, PLoS One 12 (8) (2017) 1–35.

[31] Y. Choi, J. Song, Sustainable development of abandoned mine areas using renewable energy systems: a case study of the photovoltaic potential assessment at the tailings dam of abandoned Sangdong mine, Korea, Sustainability 8 (2016) 1320–1332.

[32] J. Song, Y. Choi, Analysis of the potential for use of floating photovoltaic systems on mine pit lakes: case study at the Ssangyong open-pit limestone mine in Korea, Energies 9 (2016) 102–115.

[33] K. Trapani, D. Millar, Floating photovoltaic arrays to power the mining industry: a case study for the McFaulds lake (Ring of Fire), Sustainable Energy (2015).

[34] M. Leporini, B. Marchetti, F. Corvaro, F. Polonara, Reconversion of off-shore oil and gas platforms in renewable energy sites production: assesment of different scenarios, Renewable Energy 135 (2019) 1121–1132.

[35] S. Casini, G.M. Cazzaniga, M. Rosa-Clot, Floating PV plant and water, Chemistry, Research & Development in Material Science 7 (2) (2018).

[36] B. Castellani, S. Rinaldi, E. Bonamante, A. Niccolini, F. Rossi, F. Cotana, Carbon and energy footprint of the hydrate-based biogas upgrading process integrated integrated with CO2 valorization, Science Total Environment 615 (2018) 404–411.

[37] A. Pringle, R. Handler, J. Pearce, Aquavoltaics: synergies for dual use of water area for solar photovoltaic electricity generation and aquaculture, Renewable and Sustainable Energy Reviews 80 (2017) 572–584.

[38] I. Wasthage, Optimization O Floating PV Systems; Case Study for a Shrimp Farm in Thailand, Mälardalen University, 2017.

[39] T. Taquet, M. Dempster, Fish aggregation device (FAD) research: gaps in current knowledge and future directions for ecological studies, Reviews in Fish Biology and Fisheries 14 (1) (2004) 21–42.

[40] G. DaSilva, D. Castelo Branco, Is Floating Photovoltaic Better than Conventional Photovoltaic? Assessing Environmental Impacts, International Association for Impact Assesment, Rio de Janeiro, 2018.

[41] A. McKay, "Floatovoltaics Quantifying the Benefits of a Hydro-Solar Power Fusion, Pomona College, Claremont, California, 2013.

CHAPTER 10

Levelized Cost of Energy (LCOE) Analysis

MARCO ROSA-CLOT • GIUSEPPE MARCO TINA

1. INTRODUCTION

The current electric energy prices for industry in EU ranges from $80 to $310 per MWh, see Fig. 10.1, and are strongly influenced by taxes.

Even in other countries worldwide, we can see that the MWh price is strongly influenced by political choices and ranges from a few tens of dollars, for example in the Emirates or Saudi Arabia ($30−$40), to very high values in some islands or insulated countries (Virgin Island, $500) [1].

On the other hand, many analysts claim that the MWh price produced with solar PV is going down quickly.

The plot in Fig. 10.2 shows the best values for the MWh cost produced with PV modules in the last 5 years [2], where the last value of $19.7 per MWh has been reached in Mexico in a very large plant with horizontal axis tracking.

But what are the real average energy price and the cost production of the electric MWh?

The analysis is quite complex, and several centers of statistics and research companies give an evaluation of different techniques for producing electric energy with the relative cost [3−7].

The results of the Fraunhofer institute are summarized in Fig. 10.3 [8]. This plot has been modified as compared with the original one since the last bar on the right has been substituted by our estimate for the floating PV (FPV) plant levelized cost of electricity (LCOE).

In this plot, we suggest that the FPV technology is competitive with other RES and that the final cost is slightly lower than that for land-based PV plants.

However, the technology is at its very beginning, since 1 GWp only has been installed up to now, and we can suppose that a very large scale production would further reduce the cost of the MWh.

The FPV result quoted in Fig. 10.4 is derived from the cost of several existing FPV plants. This is taken from Ref. [9] and compared with the minimum cost of land-based PV, given by the red line 0.70 USD/Wp, and also to gable solution (discussed in Chapter 4) given by the yellow bar 0.55−0.65 USD/Wp.

The claim that FPV can be cheaper than standard land-based PV has to be justified and depends of course on the technology [10]. In the following, this point will be analyzed in detail.

2. THE COSTS OF MWH PRODUCED BY FLOATING SOLAR PV

The method proposed here can be applied to the different technologies discussed in Chapter 4 and in particular to class 1-2-3 floating plants and is based on an analysis of the different components for the plants: high-density polyethylene (HDPE), galvanized steel or other metals, PV modules, electric part (cable, inverter transformer, switches, generally called CIT).

The basic components costs assumed in our analysis are the following and have trends which are quite clear: PV modules and electric part will further decrease in the next few years, galvanized steel can be assumed approximately constant, whereas HDPE which is directly related to the oil barrel cost should probably increase in the near future (Table 10.1).

The cost obtained for the different solutions for FPV plants will be compared with that of land-based PV plants.

In the latter case, we assume that the components are the same, but we add the land preparation which is quite expensive, and we assume a cost of $80.000 per MWp (60.000 for materials and 20.000 for work). This is an average cost that in some specific cases (desert or arid zone) can be lower but that in equatorial zone can be higher due to the necessity to control vegetable growth.

Floating PV Plants. https://doi.org/10.1016/B978-0-12-817061-8.00010-5

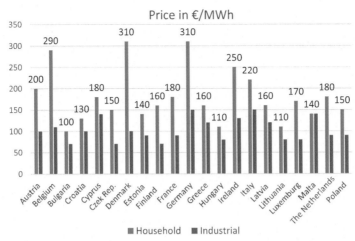

FIG. 10.1 Price in $/MWh for the main countries of EU-28 in 2018.

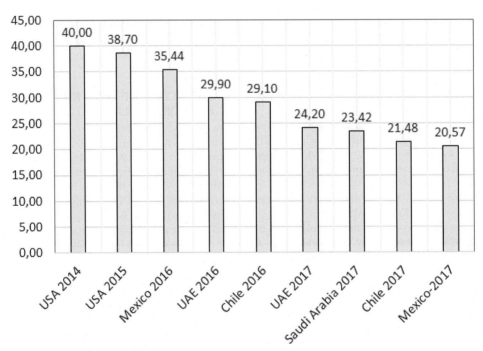

FIG. 10.2 Global solar price deflation for PV plants: cost in $/MWh. (Press Articles, IEEFA estimates.)

Raft price is more complex and should be evaluated for the different models; however, for the sake of definiteness the analysis is concentrated on the solutions proposed by Upsolar where exact numbers and finite prices are well established.

The analysis is done for 1-MWp FPV plant. We divide the costs into four components:

1. The structure which depends on technical solutions and includes all that is necessary for the PV modules support

2. The PV modules

3. The electric parts (inverters, transformer, cables, switches), without the grid connection which depends on other variables such as the grid tension and the distance of the connection cabinet.

4. Design and construction: project, assembling, Engineering, Procurement & Construction (EPC).

Raft cost depends on the number of modules per raft and on the raft structure. Three solutions are discussed:

a) Rafts with 20 modules per raft and tilt 5° (125 raft for 1 MWp). This solution was adopted in Singapore in the SERIS test bed and has been under test for 2 years.

b) Gable slender solution with 24 modules per raft and tilt 10° with 104 rafts for 1 MWp.

c) Gable2 solution with 18 modules per raft, tilt 20°, suitable for latitude above 30° and for systems with vertical axis tracking, with 140 rafts for 1 MWp.

Main results are collected for the three cases in Tables 10.2–10.5. The last column in these tables gives

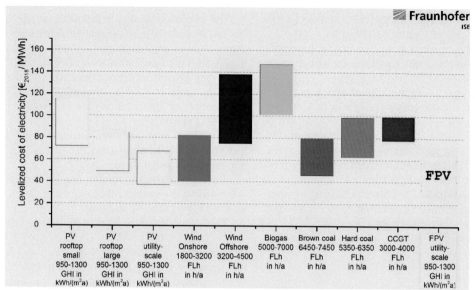

FIG. 10.3 Levelized cost of electricity of renewable energy technologies and conventional power plants. The values on the bottom line under the graph refer, in the case of PV, to the global horizontal irradiance (GHI) in MWh/(m²a), and for the other technologies to the annual full load hours (FLH).

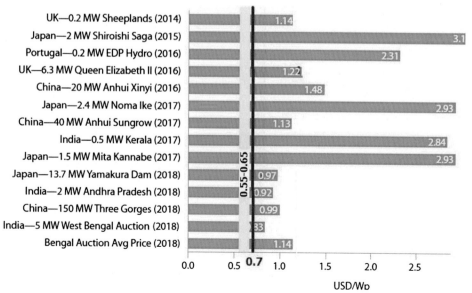

FIG. 10.4 Graph shows the cost of floating photovoltaic plants (blue line) per Watt power installed. For comparison, we have added the cost of land-based PV (red line) and the cost of our gable project (yellow bar).

the relative weight of the different components of the cost. It is important to remark that the raft cost goes down from $2426 to $982 and to $996 in the three cases discussed.

The first one is given in Fig. 10.5. The 20 PV modules of power 400 Watt installed on each raft generate a cost of the structure of $303 per kWp. This cost is quite high and not competitive with that of a land-based plant.

We note that in the raft cost 43% is due to the carpentry part and 31% to the HDPE pipes.

The second example is our gable slender proposal. The gable slender structure simplifies the raft concept and reduces the costs.

We assume that 24 PV modules of 400 W each can be installed on the raft and we get a cost of $102 per kWp only (Fig. 10.6).

In this solution, the raft cost goes down mainly because of the cost reduction of the HDPE pipes. This is due to the specific solution which reduces by four the number of pipes for rafts since a long string of N coupled raft needs only N + 1 pipes as floating support. Even if of larger diameter this strongly reduces the cost of the part dependent directly on the oil price.

The cost reduction is remarkable and raft price is more than halved.

Gable2 solution has also been discussed in Chapter 4 and is the solution we suggest for high latitudes (above 30°) and for systems with vertical axis tracking (Fig. 10.7).

In this case, 18 modules of 400 W each are installed on the raft and we get a cost of the raft for $138 per kWp. Solution with 24 modules in landscape configuration (6 × 4) is under study.

The important feature is that with the evolution of the raft project, the relative cost of the structures is going below 30% of the full plant.

For the land-based plants, we use data coming from the many existing structures, and we find the results given in Table 10.5 where we have assumed that the preparation of the location has an average price of $80.000 per MWp.

Assembly costs have been assumed to be $45.000 per MWp for all the approaches even if small differences are present between them.

The same can be said for mooring and cooling which are essential the same for Singapore and Gable raft but slightly higher for VAT tracking.

These simplifications have been done in order to go to the heart of the problem. Ground-based plants are very cheap but the floating PV systems have essentially the same cost if using gablelike structures (Table 10.6).

We should be optimistic about the possible evolution of this cost. FPV is at its very beginning, and new cheaper and simpler solutions are in evolution. Furthermore, there are not yet true economies of scale so that we can assume that the raft cost will be further decreased.

In Fig. 10.8, the share of the cost between different components of the PV plants is shown.

In standard ground-based plants, at present, the electric part and PV modules constitute 59% of the investment. In the first FPV plants, the weight of the floating

TABLE 10.1
Costs of Components.

	Costs	
PV modules	$0.25	$/W
Cables inverters, electric part	$0.12	$/W
Galvanized steel	$2.20	$/kg
High-density polyethylene	$2.40	$/kg

TABLE 10.2
Singapore Solution: 20 Modules per Raft, 125 Raft per MWp.

1 MWP Plant Cost		Relative Weight		
Modules	$250.000	31%	Raft cost	$2.426
Inverters	$120.000	15%	$/kWp	$303.2
Structures	$343.200	43%	$/PV module	$121.3
Design-assembly	$90.492	11%		
Total	$803.692			

TABLE 10.3
Gable "Slender" Solution 24 Modules per Raft, 104 Raft per MWp.

1 MWP Plant cost		Relative weight		
Modules	$250.000	42%	Raft cost	$982.00
Inverters	$120.000	20%	$/kWp	$102.3
Structures	$142.128	24%	$/PV	$40.92
Design-assembly	$78.428	13%		
Total	$590.556			

TABLE 10.4
Gable2 Solution 18 Modules per Raft, 140 Raft per MWp.

1 MWP Plant Cost		Relative weight		
Modules	$250.000	40%	Raft cost	$996
Inverters	$120.000	19%	$/kWp	$138.3
Structures	$179.440	28%	$/PV	$55.33
Design-assembly	$80.666	13%		
Total	$630.106			

TABLE 10.5
Land-Based Plants.

1 MWP Plant Cost		Relative weight		
Modules	$250.000	33%		
Inverters	$120.000	37%	$/kWp	$225
Structures	$180.000	18%	$/PV	$90
Design-assembly	$80.700	12%		
Total	$630.700			

structure was relatively important compressing the electric part to 46% and with a large part due to the rafts.

Today with new solutions the structure part of the investment is reduced and PV modules and inverters account between 59% and 62% of the full plant. This means that the decrease of cost of PV modules as well as that of the inverters will drive a further reduction of the FPV plants as well as of the MWh cost.

This, however, is not yet the electricity cost which is determined also by other parameters.

3. LCOE ANALYSIS AND MWH COST

Actually the MWh cost depends also on the cost of the investment, on the degradation of the system, the maintenance costs, and of course the radiation yield. See the

FIG. 10.5 Raft with 20 PV modules (Singapore Test Bed).

FIG. 10.6 Raft in the gable slender configuration.

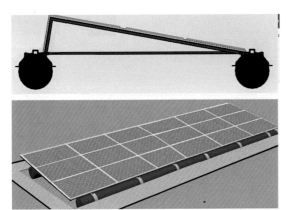
FIG. 10.7 Gable2 solution.

very detailed analysis in Ref. [11], where aspects of thermal drift, energy degradation, and integration with other energy sources are discussed.

In the following, we will perform a simplified analysis with a few fundamental inputs:

- Investment
- Financial parameters
- Energy production
- Maintenance cost

Energy Yield of the System
The energy yield is given in kWh per year and depends on the location and in turns on the latitude, on the average weather conditions, and of course on the tilt and orientation of PV modules.

In order to calculate this, we use PVGIS instrument. Table 10.7 will give the value for the Dubai.

We have assumed an advantage of 12% due to the cooling system.

Furthermore, we insist on the cooling effect as an important protection for the solar PV modules against aging.

Annual Cost
In our simplified approach, we imagine that during the life of the plant the investors have to pay a fixed loan which is determined by the initial investment and by two parameters: interest rate and number of years of loan. This value can be calculated from simple equations which can also be found directly as Excel functions.

This is not the only yearly expense. We have to add also the maintenance costs which are supposed to be constant along all the life cycle of the plant. These costs are limited for the floating plant (with some increase if a tracking system is implemented) but are on average higher for a land-based plant because of the need of maintaining the land condition in a suitable way and of periodically cleaning the PV modules (the latter is strongly reduced if cooling system is active in FPV).

We neglect here the decommissioning which, however, has an important role and which is much cheaper for floating PV where no fix structure exists, except for the mooring blocks which can be easily displaced, being fixed by a chain.

Results of this exercise are condensed in Table 10.8, where the loan and maintenance costs are given bringing to a final annual cost which divided by the energy yield gives the MWh cost for Dubai assuming a Life-Cycle Assessment (LCA) of 20 years.

The MWh price for horizontal axis tracking system has not been inserted since we have not realized any of these systems yet.

As evident from these results, the cost for floating solution can be considerably lower compared with the conventional land-based ones.

4. COST FOR DIFFERENT LOCATIONS
The result is that the gain in the MWh cost for floating plants compared with land-based plants is not negligible.

**TABLE 10.6
Comparison of the Costs for one MWp Plant.**

1 MWp Plant	Ground Based	Singapore + Cooling 20 PV 5° Rafts = 125	VAT Track + Cooling 18 PV 20° Rafts = 140	Gable + Cooling 24 PV 10° Rafts = 104
Modules	$250.000	$250.000	$250.000	$250.000
Inverters	$120.000	$120.000	$120.000	$120.000
Structures	$180.000	$343.200	$179.440	$142.128
Design-assembly	$80.700	$90.492	$80.666	$78.428
Total	$630.700	$803.692	$630.106	$590.556

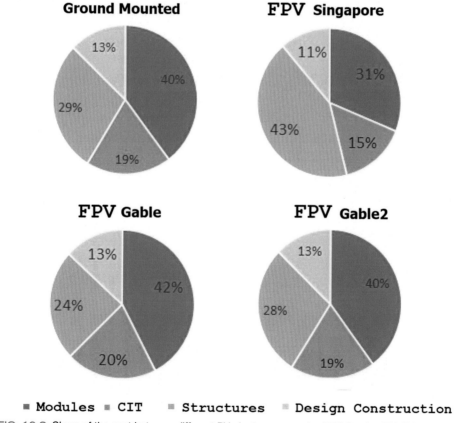

FIG. 10.8 Share of the cost between different PV plant components. *FPV*, floating PV; *PV*, photovoltaic.

In Fig. 10.9, the result of the MWh production cost with floating PV is given for 27 locations at different latitudes in the Northern hemisphere.

Calculations have been performed using the approach described above, and the gable solution has been adopted for FPV plants below 30° latitude.

TABLE 10.7
Solar Radiation Yield for Dubai (Latitude 25.07° Long 55.89°).

Solar Radiation in Dubai in kWh/y/kWp	Radiation Yield, kWh/kW/y	Cooling, +12%	Degradation 0.5%/y, −4.6%
Fix 10°	1733	1940	1851
Fix 20°	1764	1970	1885
Gable 10°	1630	1825	1741
Vertical axis tracking tilt 20°	2090	2340	2233
Horizontal axis tracking	2250	2520	2404

TABLE 10.8
Investment Cost and Final MWh Cost for FPV in Dubai.

	Investment	Rate	LCA Year	Loan	Maintenance	Energy Yield, kWh/y/kWp	Cost, $/MWh
Fix 10° + cooling	$798.392	2%	20	$48.827	$10.000	1851	$31.8
Fix 20° land based	$625.400	2%	20	$38.247	$20.000	1606	$36.3
Gable 10° + cooling	$590.556	2%	20	$36.116	$10.000	1741	$26.5
Vertical axis tracking tilt 20°	$656.606	2%	20	$40.156	$20.000	2233	$26.9

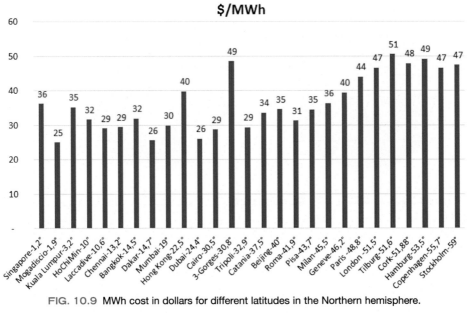

FIG. 10.9 MWh cost in dollars for different latitudes in the Northern hemisphere.

It is remarkable that there are strong fluctuations due mainly to the weather conditions. As an example, the difference between Singapore and Mogadiscio which are almost at the same latitude is mainly due to the presence of monsoon in the Singapore area which quenches the yearly energy harvesting.

Analogous plots can be produced using a system with tracking, and this would bring to a further reduction of final energy cost which could range between 20% and 30% depending on the latitude and on the solution chosen, Vertical axis tracking or Horizontal axis tracking.

The main unsolved problem in these plants is the fact that the intermittency in production is an Achilles heel, which can be overcome only with a suitable economic storage system or integration with a hybrid system such as coupling between hydro and FPV plants.

REFERENCES

[1] Electricity Pricing [Online]. Available: https://en. wikipedia.org/wiki/Electricity_pricing.
[2] T. Buckley, K. Shah, Solar Is Driving a Solar Shift in Electricity Markets, IEFEA, 2018.
[3] J. Slavic, Economics of Solar PV Power plants.an Overview of Floatovoltaic Systems' Financing, Turku School of Economics, Turku, Finland, 2018.
[4] B. Beetz, PV MAgazine: 14 PV trends for 2019 [Online]. Available: https://www.pv-magazine.com/2018/12/31/14-pv-trends-for-2019/.
[5] A. Jager-Waldau, PV Status Report 2018, JRC Science for Policy Report, Luxemburg, 2018.
[6] US Energy Information. Administration, Levelized cost and levelized avoided cost of new generation resources, Annual Energy Outlook (2019).
[7] IRENA, Renewable Power Generation Costs in 2017, Abu Dhabi, 2018.
[8] C. Kost, S. Shammugam, V. Julch, H.T. Nguyen, T. Schlegl, Levelized Cost of Electricity Renewable Energy Technologies, Fraunhofer ISE, March 2018.
[9] Where Sun Meets Water: Floating Solar Market Report, World Bank Group and SERIS, Singapore, 2018.
[10] M. Barbuscia, Economic Viability of Floating Fotovoltaic Energy, University of Lisbon, 2018.
[11] P. Campana, L. Wasthage, W. Nookuea, Y. Tan, Optimization and assessment of floating and floating-tracking PV systems integrated in on- and off-grid hybrid energy systems, Solar Energy 177 (2019) 782−795.

Index

Note: Page numbers followed by "f" indicate figures and "t" indicates tables.

Printed in the United States
By Bookmasters